Peter Bamfield
**Research and Development
in the Chemical
and Pharmaceutical Industry**

Related Titles

Bhagwati, K.

Safety Management

For Executives and Managers

2006
ISBN 3-527-31583-7

Schütz, H., Wiedemann, P. M., Hennings, W., Mertens, J., Clauberg, M.

Comparative Risk Assessment

Concepts, Problems and Applications

2006
ISBN 3-527-31667-1

Budde, F., Felcht, U.-H., Frankemölle, H. (eds.)

Value Creation

Strategies for the Chemical Industry

2006
ISBN 3-527-31266-8

Peter Bamfield

Research and Development
in the Chemical
and Pharmaceutical Industry

Third, Completely Revised and Enlarged Edition

**WILEY-
VCH**

WILEY-VCH Verlag GmbH & Co.

The Author
Dr. Peter Bamfield
56 Plymouth Road
Penarth
South Glamorgan CF64 3DJ
United Kingdom

■ This book was carefully produced. Never-theless author and publisher do not warrant the information contained therein to be free of errors. Readers are advised to keep in mind that statements, data, illustrations, procedural details or other items may inadvertently be inaccurate.

Library of Congress Card No.: applied for

British Library Cataloguing-in-Publication Data
A catalogue record for this book is available from the British Library.

Bibliographic information published by Die Deutsche Bibliothek
Die Deutsche Bibliothek lists this publica-tion in the Deutsche Nationalbibliografie; detailed bibliographic data is available in the Internet at <http://dnb.ddb.de>

© 2006 WILEY-VCH Verlag GmbH & Co. KGaA, Weinheim

Printed in the Federal Republic of Germany.

Printed on acid-free paper

Typesetting Typomedia GmbH, Ostfildern
Printing Strauss GmbH, Mörlenbach
Bookbinding Litges & Dopf Buchbinderei GmbH, Heppenheim

ISBN-13: 978-3-527-31775-2
ISBN-10: 3-527-31775-9

Contents

Research and Development in the Chemical and Pharmaceutical Industry, Third Edition. Peter Bamfield
Copyright © 2006 WILEY-VCH Verlag GmbH & Co. KGaA, Weinheim
ISBN: 3-527-31775-9

Preface

In the Preface to the first edition of this book in 1996 I expressed the hope that the contents would act as a practical guide to actions that need to be taken in the day-to-day management of R&D in the chemical industry; a "nuts and bolts of management" as accurately described by one reviewer. That this proved to be the case was evidenced buy the fact that a second edition was published in 2003, in which the contents were expanded and modified to include management issues relevant to both the chemical and pharmaceutical industries.

Way back in 1996 little did expect that some ten years later I would be asked to write a third edition, but as can be seen this has proved to be the case. I have therefore taken the opportunity to bring up to date relevant sections on organisation of R&D in a global context, outsourcing and virtual companies; laboratory automation, regulatory affairs and product registration; project, change and knowledge managements to reflect current practice in these areas. I wish to thank my good friend Peter Murray for his help with the latter two topics.

The style and grammar of the first two editions of this book were greatly influenced by the constructive criticism I received from my late wife Mary, in whose memory I wish to dedicate the new edition.

Peter Bamfield, 2006

Research and Development in the Chemical and Pharmaceutical Industry, Third Edition. Peter Bamfield
Copyright © 2006 WILEY-VCH Verlag GmbH & Co. KGaA, Weinheim
ISBN: 3-527-31775-9

Introduction

Research and Development in the Chemical and Pharmaceutical Industry, Third Edition. Peter Bamfield
Copyright © 2006 WILEY-VCH Verlag GmbH & Co. KGaA, Weinheim
ISBN: 3-527-31775-9

1
An Overview of the Scope and Contents of the Book

1.1
The Industry

Chemistry, due to innovations commercialised by the chemical and allied industries, has a long history of making major contributions to the welfare and development of mankind. Without the myriad of industrial applications of chemistry we would all be so much the poorer, especially in the developed world. Our health would be far worse than is nowadays imaginable and our lifespan much shorter due to untreatable diseases and poor methods of diagnosis. Food would be much less available and with less choice, whilst the clothing we wear would be less functional and aesthetically pleasing. Energy would be more costly and inferior, rendering all sorts of activities that today we take for granted, especially in our homes, non-viable. Rafts of new materials which have lead to new construction methods for household goods, buildings, and the means of transportation such as cars and aeroplanes, would not be available. Last but not least we would not have witnessed the immense changes in our means of communication, both audio and visual, that have occurred in the last 25 years.

However, if a group of people were asked for a definition of the chemical Industry there is no doubt that most would give answers that would not reflect its massive historical contribution to the well being of society. It is a paradox that chemistry and chemists are perceived by the public as good but that chemicals and hence the chemical industry as bad. This poor image is largely due to a legacy of poor environmental control by the industry that existed until the latter part of the 20th century, which has lead to continuing attacks even today from the environmental lobby. Ironically major improvements in the environment will come mostly from the innovation and drive, which is at the core of R&D within the Industry.

Against this background in the recent past, not surprisingly, many sectors of the industry sought to distance themselves from the main stream, not only for social and political reasons but also for a perceived economic advantage. For instance, those chemical companies which had a strong base in medicinal chemistry, demerged and aligned themselves with the biological sciences to become part of the international Pharmaceutical Industry. Similarly those based on developing technologies have wanted to be seen as entirely new industries e.g. the biotech industry. In this way these sectors hoped to be viewed, in the eyes of the general public and investors, as cleaner

Research and Development in the Chemical and Pharmaceutical Industry, Third Edition. Peter Bamfield
Copyright © 2006 WILEY-VCH Verlag GmbH & Co. KGaA, Weinheim
ISBN: 3-527-31775-9

than the general Chemical Industry, making them more attractive to the world stock markets. However, even here there are hidden pitfalls, as has been demonstrated by the continuing debate over pesticides from the agrochemicals sector, the ongoing furore over GM crops, nanotechnology and "grey goo", whilst "big pharma" took a hit from its perceived behaviour in the third world as portayed in the the film The Constant Gardner, and accusations of overchanging for drugs in the US [1-6].

In spite of all these changes in this book, the term Chemical Industry will often be used in its broadest possible context, covering the pharmaceutical business, even if not specifically stated in parts of the text. The content will therefore be of relevance to all those professionals who are already working, or new recruits about to start, in the R&D function of any company which uses both chemical and biochemical processes to invent and produce products or services for commercial gain. This will

Table I1   Chemical and Allied Industries

Generic Groupings Type	Example
Raw and Bulk Materials	Petrochemicals
	Inorganic
	Minerals/ores
	Ceramics/Glass
	Detergents
	Surfactants
Fine Chemicals	Organic Intermediates
	Fine Inorganics
Specialities	Photographic Chemicals
	Dyes and Pigments
	Perfumes and Fragrances
	Adhesives
	Coatings
	Food Additives
	Electronic Chemicals
	Biocides
	Nanotechnology
Biological	Pharmaceuticals
	Agrochemicals
	Biotechnology
	Food Products
Polymers	Fibres
	Plastics
	Performance Materials
Services	Contract Manufacture
	Analytical
	Toxicology

encompass the raw material and bulk chemical producers, the manufacturers of fine chemicals and intermediates, the speciality and performance chemicals industries, the chemical biology science based industries such as pharmaceuticals, agrochemicals and food products, as well as those operating in the polymers and materials chemistry areas, and the providers of outsourced services e.g. contract research, contract manufacture, analysis, toxicology, efficacy testing and safety. These generic groupings are shown schematically, but not comprehensively, in Table I1.

There are several industries that employ many chemists, which are not normally classified as part of the Chemical Industry, e.g. Energy, Water, Food and Agriculture, and this book should be of value to the R&D Managers operating in these areas. Additionally the book will be relevant to those working within the research institutes, public health and trade association laboratories, where the working environment is essentially the same as within industrial R&D.

In the second edition of this book in 2003 two major activities, that had changed the structure of the industry since the first edition in 1996, were outlined; namely mergers and acquisitions and globalisation. These acitivities have continued without any lack of pace to the present day.

Globalisation, whilst nothing new for the chemical industry, has become an economic necessity in many sectors and is more often accompanied by some form of merger or acquisition. A classic case is that of the European colorant manufacturing industry [I-1]. At the beginning of the 1990s there were six major European companies involved in the dyes and pigment industries: BASF, Bayer, Ciba, Hoechst, ICI and Sandoz. Following a series of mergers and demergers, there are now only three companies: Ciba, Clariant and DyStar. The progress from the three to six is charted in Table I2.

The pharmaceutical industry has seen even more dramatic changes with mergers ands acquisitions on a mega scale. The origins of the some of the current, largest pharmaceutical companies are shown in Table I3 [I-6]

These changes, in what were formerly household names and symbols of the greater Chemical Industry, have been matched by dramatic changes in individual companies as exemplified by ICI in the UK. Some of the changes that occurred to this company during the 1990s are shown graphically in Figure I1 [I-2, I-3]. Similar changes have happened and are continuing in such former bastions as BASF and DuPont.

All these changes within the industry have implications for the R&D environment in which managers have to work. For instance, it is not uncommon for research

Table I2 Restructuring of European Dyes Industry

Year	Merger Activity
1993	ICI demerges its dyes business forming Zeneca
1995	Bayer and Hoechst merge dyes businesses to form DyStar
1995	Sandoz demerge dyes as part of Clariant
1996	BASF acquires dyes business from Zeneca
1997	Ciba forms Novartis, dyes go into demerged Ciba Specialty Chemicals
2000	BASF transfers dyes business into DyStar
2004	DyStar purchased by Platinum Equity

groups to be working on the same topic in China, India, Europe and the USA, enabled by rapid electronic communication systems that provide information on a 24 hourly basis. Management is most likely to be in a matrix, which operates across functional, cultural and time barriers and hence requiring a high level of coordination and different interpersonal skills from those used in a small, local R&D group. In other instances the Manager may be working in a company providing R&D to another company under an outsourcing contract, placing a different emphasis on certain of these interpersonal skills, such as negotiation. In this global environment the management of change is crucial.

1.2
The Role and Breadth of R&D

In this book the role of R&D in the Industry is specifically defined as follows:

> *The invention and development of products, processes, systems and services which will provide the company with a commercial opportunity.*

More broadly its role can be described as providing options, to colleagues in Marketing, for the potential growth of the business. In other words, R&D provides the opportunity whilst Marketing is charged with making the decision on whether to pursue this opportunity, and then with its commercial exploitation.

With pure or fundamental research the provision of commercial opportunities is, by its very nature, a long way into the future. However, in industry this is the final objective, no matter how far into the future, it is not science for the sake of science. This type of long-term work is uncommon in most chemical companies and at best only represents a very small fraction of total R&D budgets, even in the largest companies. It is often contracted out or carried out in collaboration with research institutes and universities, these bodies finding that it provides them with a useful source of funding and indeed a new role in the economic welfare of a country. The time frames and measures of success are quite different from product oriented research and when carried out in industry requires special management skills. This will be dealt with in Section C on innovation and creativity.

Table I3 Mergers of Pharmaceutical Companies

Original Companies	Years	New Company
Pharmacia, Upjohn , Monsanto, Warner Lambert, Pfizer,	1995, 1999, 2000, 2003	Pfizer Inc
Wellcome, SmithKline Beecham, Glaxo	1995, 2000	GlaxoSmithKline
Astra AB, Zeneca	1998–9	AstraZeneca
Hoechst, Roussel, Marion Merrel Dow, Rhone-Poulenc	1994, 1995, 1999	Aventis
Sandoz , Ciba Geigy	1996	Novartis
Squibb, Breistol-Myers	1989	Bristol-Myers Squibb

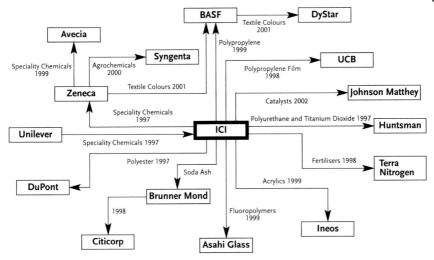

Figure 11. Change of Ownership of ICI Business 1993–2002

The heartland of innovation research, the R of R&D, in the Industry is the search for new products. These products will help to differentiate the company's product range from that of the competition, thus giving it a competitive edge in the market place. In order for this type of work to be done effectively by R&D it is very important to set clear targets. An R&D Manager must have an understanding of the particular market where the company operates, and especially recognise its driving forces. The R&D Manager also needs to work closely with colleagues in Marketing during the process of selecting and evaluating R&D targets. These important steps in the development pathway, or Innovation Chain as it is often called, will be described in Section D, which deals with Project Management.

In development, the D of R&D, there is a very strong overlap with the R&D people and those working in both manufacturing and marketing. They are all part of the operational function of the company and in these circumstances the R&D Manager will be required to be commercially astute. In addition to fitting into this matrix the R&D Manager has to be competent in working within very short time frames and in meeting deadlines. The ability to respond to circumstances, which may change rapidly, is also a necessary skill. The Project Management skills described in Section D are essential assets in helping the R&D Manager carry out this task in an effective manner.

Another area of R&D is that looking at new ways of applying or using existing chemical products; this is sometimes called applications research. It is often carried out alongside a technical service role to customers. It therefore has a very strong interface with customers in the world outside the company. In fact in many small companies, especially those operating in the speciality or performance chemicals area, this type of R&D is the only sort that is carried out. This is because its time frames are short and entry costs low, since it does not carry those associated with new product introduction, such as product registration. The R&D Manager in this environment is very much a part of the commercial arm of the company.

Good quality R&D work requires an excellent back up from what are called the support functions. These include analytical, information services and technology, intellectual property and regulatory affairs. The management of these functions, especially the laboratory based analytical service, requires many of the skills associated with core R&D and will therefore not be treated separately in any detail, except to point out their relationship and role within the organisation. Traditionally, the support functions were an integral part of a large company's R&D function but nowadays they are most likely to be outsourced under contract to external organisations specialising in a particular service, especially if these activities are not seen to offer a competitive advantage to the company. The organisation of outsourcing is not a simple matter and can often take up a significant amount of management time, especially if the contract is not negotiated in an effective manner. This will be covered in Section B.

1.3
R&D Organisation

The organisation of R&D is very dependent on the size, nature and the location of the Company which it serves. It is clear that, like almost everything else in life, there is no ideal organisational system that, if only we could set it up correctly, would produce the results required in a smooth and untroubled way. Organisations are always changing, mostly for the better, but not always if the changes are ill thought out and carried out in haste. There is a dynamic which switches from some form of functional management system to self managed teams or empowered organisations over a period of time, depending on the nature of the work, the company strategy or the prevailing socio-economic climate. However, the truly monolithic functional organisations, known as "first generation research", are things of the past as we have moved through the project lead R&D of the "second generation" into the market driven "third generation" organisations aligned directly with the company strategy [1-4]. We have now moved into the "fourth generation" where the management of knowledge has become a key role for R&D [1-5]. The issues involved in the different types of structures that exist within companies, especially chemical ones, will be covered in Section B.

Whilst a knowledge of these organisational structures is essential, a first time R&D Manager is most likely to be involved in the specific task of laboratory management. Laboratories come in all shapes and sizes and are used to perform a large variety of functions. These can range from superbly equipped synthetic and analytical laboratories in a major pharmaceutical company, semi-technical laboratories with larger scale reactors used to study the scale-up of reactions in a custom manufacturer's premises, technical service laboratories supporting a specific business with specialised equipment, to a sparsely equipped laboratory in a small company or a field trial station, remote from the centre of excellence. Whatever the situation in these laboratories they require the same basic set of the management activities; employing the right people (Section A), providing the required equipment and support, controlling costs and budgets, ensuring good health and safety is practiced and that the work done is of the highest quality (Section B).

1.4
R&D Personnel

Since this book deals with the R&D functions in the chemical and pharmaceutical Industries, most, but not all, of the professional staff working in R&D will be chemists. The chemists will have been trained in the major disciplines, organic, inorganic, physical and biochemical, often having further specialised in areas such as polymers, colloid science, computational chemistry, analytical science, medicinal and bio-chemistry, photochemistry, process chemistry, etc as the move to a more interdisciplinary style of education has taken place. Other professionals will come from related scientific backgrounds, including biology, microbiology, medicine and genetics, botany and ecology, toxicology and environment, mathematics and physics. In the process development field there be a strong involvement in the management process by chemical engineers, as well as a close association with mechanical engineers, civil engineers and metallurgists.

The Industry is being increasingly organised around multi- or inter-disciplinary teams, with the edges between the various disciplines becoming blurred. Thus an R&D Manager in the modern, broadly based, multi-disciplinary, trans-national Industry is just as likely to be a chemical engineer, biochemist, biologist, microbiologist or physicist as a professionally trained chemist. Such a person will have to manage professionals from all of the other disciplines listed above, and probably others as well. In addition to the professionally qualified scientists, there will also be technical staff, often in the process of undergoing training, as well as some non-technical staff to be managed. With the arrival of the flatter organisation, career development for both the R&D Manager and staff has become more difficult, leading to a need for greater exposure activities outside the laboratory in the world of business, and further training as part of a process of continuous professional development. The human resource skills required to recruit and then manage the performance of the desired staff, together with related managerial activities will be considered in Section A.

1.5
Creativity and Innovation

Much has been written over the years on how to create the correct climate in which creativity and innovation will flourish. A great deal of this material is of an academic nature and tends to dwell on abstract themes rather than concrete ones. Section C, whilst not ignoring these "intangibles" is designed to offer practical advice on the steps which can be taken by an R&D Manager, in order to foster a creative climate and to assist the process of innovation. Included is a description of the tools which can be used to help the creative process; the idea generation which is the starting point for any innovation. The protection, by patents and other means, of the intellectual property produced by R&D, together with the management of knowledge and possible ways in which it might be exploited is also covered in Section C.

1.6
Project Management

This is an extremely important subject and over the years many detailed, erudite and complex books have been produced dealing with its application in a variety of areas, but mostly outside chemistry R&D. It is beyond the scope of this book to go into great detail on project management. The emphasis in Section D of this book is on the selection and evaluation of the R&D targets, including portfolio management, together with a description of the practical application of the principles of project management to the innovation chain. The management of time along this innovation chain is a key issue, so that the new chemical products coming out of research can be delivered to the market in the shortest possible time. The nature of the project will define the managerial skill requirements of the project leader and these will be discussed in some detail.

Overview
By the end of this book the reader should understand the principles inherent in running a successful R&D organisation in any of the sectors of the Chemical, Pharmaceutical, Chemical Biology and allied industries. This will include knowing the requirements for harnessing the human resource, organising the environment for a climate of creativity and then managing the resultant innovations through to success in the market place.

References

[I-1] BAMFIELD, P., *The Restructuring of the Colorant Manufacturing Industry.* Review of Progress in Coloration, **2001**, 1–14

[I-2] Data taken from a presentation by ALAN SMITH, AZ-Tech

[I-3] BROPHY, J., *Chemistry in Britain,* **2000**, 31(4), 45

[I-4] ROUSSEL, P.A., SAAD, K. N., ERICKSON, T.J., *Third Generation R&D; Managing the Link to Corporate Strategy,* Harvard Business School Press, Boston, **1991**

[I-5] MILLER, W.L, MORRIS, L., *Fourth Generation R&D, Managing Knowledge, Technology and Innovation,* Wiley, New York, **1999**

[I-6] ANGEL, MARCIA, *The Truth about Drug Companies,* Random House, **2005**

[I-7] ROYAL PHARMACEUTICAL SOCIETY, May **2004**, http://www.rpsgb.org.uk/pdfs/mergers.pdf

Section A
Harnessing the Human Resource

Personnel costs in an R&D department or group within the Chemical Industry often account for greater than 75 % of the budget, but more importantly the people employed in the group's activities contribute to 100 % of the output. Consequently, the single most effective managerial task for an R&D Manager is the harnessing of the human resource to the job in hand.

The topics discussed in this Section are of importance to the management of people in any environment, however, they are viewed from the unique standpoint of the creative and professional environment that exists within industrial R&D, as exemplified by the Chemical and Pharmaceutical Industries. The objective of the management of human resources is to recruit, maintain and develop staff able to carry out the activities of the R&D function or laboratory at the optimum level. To this end it is necessary to carry out the following specific activities:

- Appoint/recruit staff of the right standard and with the necessary skills
- Develop the skills and competencies of the members of the laboratory by providing effective training
- Build strong teams, working in a supportive climate, with challenging objectives
- Monitor the performance and provide appropriate rewards to staff
- Plan and anticipate future staff requirements

At the core of these activities is the need for an understanding of the skills base that is appropriate to meet the research objectives of the organisation and then matching this with the capabilities of the current personnel of the group concerned. The methodology for carrying out these requirements is dealt with at the beginning of this Section. The recruitment of new staff to fill any gaps in the skills base and the subsequent development of all these people into enthusiastic and successful chemists and scientists is then examined. The skills base of any R&D Group is a summation of skills of the individuals who make up such a grouping. Each of these individuals will have his or her own performance management and career development requirements and guidance is provided on both of these subjects. The people employed to work within R&D will be expected to work in teams, often with scientists from other disciplines or colleagues from other parts of the business as part of a multi-disciplinary team, and the skills a manager will require to effectively run these teams are considered.

Research and Development in the Chemical and Pharmaceutical Industry, Third Edition. Peter Bamfield
Copyright © 2006 WILEY-VCH Verlag GmbH & Co. KGaA, Weinheim
ISBN: 3-527-31775-9

Overview

Competency building is an essential activity so get to know what your people can do. The first priority in running a quality organisation is to employ and retain quality people; having a sound methodology is the basis for successful recruitment. Managing an individual's performance to meet the desired objectives is a necessary step in helping to ensure the company's goals are met. Most people want to see a career path stretching in front them, reassure them by having career development plans. Succession planning may be almost impossible but at least make contingencies. Team working is key to running a successful R&D Group. Leadership qualities depend on the nature of the team. Appoint charismatic people rather than technical superstars to lead creative teams. Don't forget to manage your own career.

1
Building the Scientific Skills Base of the Group

Competency building is an essential activity.

It is essential that any R&D Group demonstrates a high technical competency in its chosen area of work; otherwise it will not be able to meet the needs of the business and will be in terminal decline. Therefore, one of the key tasks for the R&D Manager is the building and maintenance of this technical competence. To carry out this task effectively the R&D Manager must first of all have a clear understanding of what skills or competencies are actually required within the R&D Group in order to meet the ambitions for innovation stated in the company strategy. These skills are those that lie at the heart of the organisation, making up its character and defining the area within which it has chosen to operate. Such technological skills are an extremely important component and are known as *Core Competencies* [A-1, A-2]. Core competencies are those skills which allow the organisation to build its existing business or to diversify into new areas, which are derived from these competencies, but which will be used in a non-traditional manner. If the business desires to move into an entirely new area it will be necessary for R&D to either build or, more likely, acquire new skills in order to have the competency needed to satisfy this ambition.

It will be unusual for an R&D Manager to be in the fortunate position of being able to build a team with the required competencies from scratch. Normally, an R&D Manager, either newly promoted or moving into a new job, will have to work with the staff that is available. Whichever is the case, to be successful in the role, a Manager must know the skills that the team currently possesses, and also those which will be required for the future to implement the strategic plan. Only once this has been done will it be possible to build and maintain competencies, and hence have the necessary *Skills Base* for the R&D Group.

The process for ensuring that the right skills are available within the R&D group involves analysing what skills the business requires, auditing what currently exists and determining if there is a gap that needs to be filled. Any gap can then be filled either by retraining of the existing staff or by transfer from elsewhere in the organisation or by external recruitment of new staff.

Research and Development in the Chemical and Pharmaceutical Industry, Third Edition. Peter Bamfield
Copyright © 2006 WILEY-VCH Verlag GmbH & Co. KGaA, Weinheim
ISBN: 3-527-31775-9

1.1
The Skills Audit Process

Get to know what your people can really do.

The systematic approach to analysing the competency of the R&D Group is the *Skills Audit*, which involves the following three steps.

1. Determination of the skills required for the R&D Group to fulfil its role, the *Skill Requirements*.
2. Auditing the current skills of the staff in the R&D Group, the *Skills Base*.
3. Determination of the difference between the skills required by the R&D Group and the skills of the currently available staff, the *Skills Gap*.
4. Without carrying out the analysis in a thorough manner, not only will a Manager lack a complete awareness of the capability of the group and its ability to carry out its strategic function but will also be unable to define the areas for improvement.

1.1.1
Determining the Skill Requirements

The process for defining the *Skill Requirements* of the group involves generating answers to the following questions:

- What is the size and nature of each project within the portfolio assigned to the group?
- What professional skills are required to carry out the specific tasks within each project?
- How many people with the specific skills will be required for the coming period?

Generating the answer to the first question is simply a matter of listing the number of R&D projects that are in the plan for the forthcoming period, the *Project Portfolio*, and then allocating the number of people to each of the projects as allowed by the budget. Any overall manpower problems should have been flagged up earlier during the budget setting process. Too few people would have indicated the need to expand the team, either by internal redeployment or by recruitment from outside the company, whilst too many people for the total programme would have presented a quite different problem. These aspects of budgetary control and target setting are dealt with in Sections B and D. For the purpose of this exercise we will assume that excess personnel is not an issue.

To answer the second question a list is made of the broad technical skills required for each project, e.g. synthetic chemist, colloid scientist or biologist, which will be needed during its lifetime. It is important at this stage not to forget any technical or other support that may well needed from outside the group or company, for instance from analytical services, computer modelling and information services.

To address the final question it is necessary to break down these broad requirements into the specific skills and the number of people with these skills that will be needed for each project. For example, the synthetic chemistry requirement could be as widely different as the synthesis of monomers for polymers or chiral intermediates for pharmaceutical or agrochemicals. It is technically inefficient to ask an expert in one of these areas to carry out work in another area. This may well be forced upon a Manager due to staff shortages, or by a short term no recruitment policy in the company, but if this happens sufficient time must be built into any project to allow for the re-focusing of a persons fundamental skills and the training that will be necessary.

In the example of a simple matrix given in Table A1, the skills requirements for the forthcoming projects are seven synthetic chemists, specifically skilled in heterocyclic synthesis and sugar chemistry; three polymer chemists, specifically with experience in water soluble polymers and conducting polymers; one colloid chemist and two microbiologists of general capabilities. Also identified are the needs for significant analytical support, modelling, and the use of semi-technical manufacturing personnel and equipment and the guidance of a physicist.

Table A1 Estimation of Skills Requirements

Project	Skills	Numbers	Specific Skills	Support
1	Synthetic Chemist	4	Heterocyclic Synthesis	Analysis Modelling
	Microbiologist	1		
2	Synthetic Chemist	3	Sugars	Analysis
3	Polymer Chemist	2	Water Soluble Polymers	Semitechnical Scale manufacture
	Microbiologist	1		
4	Polymer Chemist	1	Conducting Polymers	Analysis
	Colloid Chemist	1		Physicist

1.1.2
Auditing the Current Skills

The ability to use effectively the skills of an R&D Group is highly dependent on the Manager having an understanding of what the group already know. This may seem an obvious statement to make, but it is an area that is not given sufficient attention by many Managers. Consequently the absence of this information leads to many opportunities being missed where the skills or experience of individuals could be used to everyone's advantage. This is especially the case with those people who have gained their specialist knowledge when working outside their current roles. Their skill may have been gained within another department, from past experience with another employer or even from an involvement in outside interests or hobbies.

In small groups it may be less common for colleagues not to know each other's skills but it is still surprising how often the flow of information via the informal network falls down. The right question is not asked of reticent individuals, their talents lie buried and are therefore unused. When it comes to larger groups the problem in knowing the total Skills Base is multiplied, unless a systematic method of data collection is adopted.

The carrying out of an audit of staff is no simple matter and it is an exercise that must be well planned. Most of the explicit material about the careers and training of staff will be in staff records but a check on currency and completeness is advisable. However, it is tacit knowledge that is much more difficult to discover and record and is often the most valuable (see Section C, 3.5.1). Basically the process involves asking the person to describe their current activities and those of the recent past for explicit knowledge and then teasing out from them their tacit knowledge or high-level know-how. Physical evidence of an individual's ability in a particular skill should also be collected at this stage, e.g. workshops attended with tutor's reports or citations by known authorities or certificates from examining bodies. This methodology is known as *competency based assessment.*

The selection of the process to be used to carry out the audit will depend on the circumstances. Traditionally, this has been done by filling in a questionnaire and is still the simplest method for small groups of people. One of the problems with questionnaires is that they are often poorly designed to find out tacit knowledge and there is also a high resistance factor from the people filling in the answers or being questioned, and this may cause valuable information to be missed. Using a one to one interviewing process can eliminate this problem, whilst in addition demonstrating a personal commitment to the exercise. This process can be very time consuming with larger groups and may be one reason why it is often not done or, when it is, carried out in a rushed and unsatisfactory manner. Fortunately, computer programs have been devised which make this task much easier to perform and also to store and subsequently analyse the data, some of which are available online [A-3]. A Manager must remember that such personal data is covered by a legal framework, such as the Data Protection Act in the UK and by equivalents in other countries.

There are a great many areas in which a Manager will wish to collect data about staff but usually these fall into a few generic classes as given in Table A2.

Having carried out the audit a Manager will have a valuable inventory of the group's skills. However, this is not a once and for all exercise, the inventory must be kept up to date as the Skills Base will change over a period of time, both by the growth of the internal knowledge base and by personnel changes within the group. The cross referencing of this inventory with the individual's training records is one effective way of maintaining the currency of the data.

1.1.3
Discovering the Skills Gap

After carrying out the exercises of the analysis of the Skill Requirements and the Skills Base it is possible to discover if there is or is not a *Skills Gap* in the abilities of the group. This is arrived at by a simple difference technique on the two sets of data.

Skills Requirements – Skills Base (current) = Skills Gap

A Manager, having discovered that there is a Skills Gap, must make its removal a high priority activity. Timing is essential, especially where the missing skill is likely to be a rate-determining factor in the success of any project. Once the data on Skills Base is held on an updateable database, an early warning of such problems can be flagged and discussed during the R&D target setting meetings.

To show how the methods to derive the Skills Gap are used, let us consider the case of an R&D Manager whose job includes running a team working on bioactive materials for industrial type applications.

The plan for the coming year requires that work be done on four separate projects. The total budget that has been agreed with the Business Manager allows for nine people to be allocated across the portfolio of projects. The agreed allocation of the nine people to each project is as shown in Table A3.

Project A involves the synthesis of novel biologically active materials for testing in the preservation of the latex used in the manufacture of paint. From past experience, it is known that the chemicals that are likely to be active in this environment will be those based on heterocycles containing sulphur. Therefore for this Project the particular Skill Requirement is the ability to design and synthesise sulphur-containing heterocycles that will show bioactivity in this industrial outlet. The two people who will be required for this task are available in the cur-

Table A2 Skills Audit – Classes of Information

Class	Examples
Education	Qualifications
	Work related training
	Language skills
	General
Work experience	Current role
	Previous jobs
Expertise	Specialist
	Technical
	High Level Know-How
Publications	Internal Reports
	External Papers
	Books
	Patents
	Lectures
Outside Activities	Professional Bodies
	Learned Societies
	Trade Associations
	Relevant Contacts

Table A3 Deriving the Skills Gap

Project	Budget Numbers	Skills General	Skills Specific	Skills Available	Skills Gap	Support
A	2	Snythesis	Sulphur Heterocyclic	Most	Training required In automated synthesis	Analytical
B	4	Synthesis	Low MW Polymers	Yes		Analytical
		Process Dev.	Semi Technical	Yes		Plant Staff
		Formulation	Liquids	Yes		Statistics
		Microbiology	Field Trials	No	Microbiologist	
C	1	Formulation	Powders	Yes	None	None
D	2	Microbiology	Testing in Paint, Plastics	Yes	None	Polymer Compounding
Totals	9			8	Microbiologist	

rent staff. However, they lack the skills to use the updated automated synthesis equipment to be installed shortly in the laboratory and hence retraining will be required (see Section B). The application testing of the materials that they produce is to be provided by those allocated to work on Project D. Chemical analysis of these products will also be needed and the provision of the appropriate resource is to be agreed with the analytical manager or outsourced as necessary.

Project B is at a more advanced stage than Project A. It requires the preparation of larger quantities of the selected materials for use in field trials. Chemists are available with the necessary experience of both synthesising these types of materials and in developing processes for the scale up of their manufacture. They have recently been on a training course in the use of the new automated equipment and are familiar with the techniques of parallel synthesis. The chemical product from the semi technical scale manufacture needs to be formulated into a liquid that is suitable for application in the field. The person with the required skills is available from within the current team. However, there is a problem when it comes to carrying out the field trials. These trials will need to be supervised by a microbiologist with the appropriate experience of operating in the field. At the present time there is no one available who has done this, either in the recent past or who could be trained to do so on any reasonable time scale to met the needs of the current Project. A specific Skill Gap has therefore appeared and, since it is a skill that is crucial to the success of the Project, requires prompt managerial attention to rectify the position. More extensive external support is also required for this Project, not only analysis but also from the team operating on the manufacturing plant. Additionally, someone skilled in statistical experimental design will be required to help set up and help in the analysis of the results from the field trials. The handling of a wide-ranging project like this one, which will be team orientated, is covered in Section D.

Project C, the formulation of actives into powders presents no problems with regards to staffing, the single person who is required is sub contracted from the larger, core formulations group. There are no difficulties with Project D, which will provide the application testing support for all of the other Projects. However, Project D will involve extensive testing of the bioactive materials in polymer films and the provision of support in the compounding and film blowing needs to be organised from the polymer research group.

In summary, as shown in Table A2, Gaps in the required skills have been identified. Most gaps can be filled by the retraining of existing staff or by outsourced support (see Section B) but the requirement for a microbiologist who is experienced in carrying out field trials cannot be filled in either of these ways. This skill is essential to the success of the Project and urgent steps must be taken to fill this Gap.

From this example it can be seen that the data allows the training needs of existing staff to be defined and hence retraining to be carried out, often as part of the performance management process (see Section A, 2.1). Additionally, having this data makes it much easier for a Manager to define the selection criteria for new recruits and to structure the interviewing process. The guidelines in the process for recruiting professional and technical people will now be discussed.

1.2
Recruitment

The first priority in running a quality organisation is to employ quality people.

As has ready been stated the main component of any successful R&D group is high quality staff. Therefore, one of the most significant duties a Manager will have to perform is the recruitment of personnel, either to expand the group or to replace people who leave for a variety of reasons. This is a key element that will be used in the judgement of a Manager's personal performance. Recruitment is such an important activity that Managers must make every endeavour to ensure that they are directly involved with the process, not only when people are being recruited to work immediately in their groups but also if they are likely to do so in the future. In this Section the emphasis will be on the recruitment of professionally qualified staff into R&D. The principles described are applicable to any professional group whether they be chemists, chemical engineers, biochemists, microbiologists or any of the others who are employed in the modern Chemical, Pharmaceutical and Allied Industries. The recruitment of the support staff, especially at the technician level, follows the same general principles, modified to take account of their more limited education or job experience, and the nature of the tasks they will be required to perform in R&D.

There is a strongly held view that R&D personnel are different in nature from other employees, requiring special methods to be applied when they are being recruited. Certainly, R&D people do constitute a definable group and do display some characteristic personality traits, but some of these may be related to the culture in which they have developed. Whilst it is dangerous to generalise about individuals, especially creative scientists, the following characteristics are discernible within a broad group of R&D personnel.

- Independently minded. Creative people especially like to challenge the accepted norms and to seek alternatives.
- Highly motivated. They have a strong desire to overcome problems and to achieve a goal, are self-disciplined, focused and flexible thinkers.
- Self-effacing. They show a tendency to underplay their abilities even though having confidence in themselves and being self-aware.
- Introverted. They are often less interested in people and more in ideas and concepts.

These characteristics, which are often contradictory with each other, must be taken into account when considering what type of person is to be recruited. These personal characteristics are therefore amongst those that will constitute the key criteria that are used during the interviewing process. The interviews should be structured to overcome any problems which might be associated with, for instance, self effacement or introverted behaviour on the part of the candidates, and allowance should be made for the independent mind which could, if not handled properly, just seem to be perverse. However, in spite of these differences, the basic recruitment methodology to be used for R&D personnel is the same as for any other group of staff within the company [A-4].

Recruitment is a time consuming and expensive activity and therefore needs to be carefully planned. Errors made at the recruitment stage can result in staff and personal problems for Managers for years to come. There are many packages produced by management consultants that can be used to assist in the process of staff selection. It is quite likely that a company will use one of these and any new Manager will have to be trained in their use, as part of the personal development plan, before being involved in recruiting. However, Managers should not think that these packages are foolproof or that they contain a magic formula, which, if applied rigidly, will guarantee success. The packages will all have many common features and these are described in this book. The ability to recruit effectively will, like all skills, only be improved by regular practice, but a sound method is a good basis on which to build these skills. Time that is devoted to learning and practising these skills, either on a training course or by sitting in with colleagues during their recruitment activities, is time well spent.

The key elements in the recruitment process are: having an accurate description of the position to be filled together with the technical and personal skill requirements of the job holder, the placement of the advertisement, the selection of the candidates for interview, the interviewing process and selection methodology and finally the making of the offer which leads to acceptance. This is shown schematically in Figure A1. If carried out in the sequential manner shown in Figure A1, the elapsed time for the process would be immeasurably long, consequently several activities should be executed in a parallel process.

There is an alternative to carrying out this long process and that is to employ "headhunters". These are search consultants who will either get a specific candidate for a job or, more commonly, provide a short list of people to be interviewed by a Manager and colleagues. This method of recruitment is used primarily when looking to fill the more senior, executive jobs or when a very specialised skill is required [A-5]. However, there is a very effective form of headhunting that can be performed by all R&D Managers. This is when graduates, especially postgraduates, are being recruited directly from university. It involves direct contact with their supervisors in the chemistry departments of these establishments, seeking the names of high quality candidates who can become targets for the company's advertising at the time they are about to graduate. This method will only work, and be a reasonable activity, if the groundwork of building good relationships with the academic staff in the chemistry and related departments of universities has been done over several years.

1.2.1
Job Definition and Evaluation

Most sizeable companies will, as part of their Quality and/or Performance Management Systems, have descriptions for all the jobs that are being undertaken within the organisation. The production and use of these in the context of the management of performance will be covered in greater detail in Section A, 2.1. If the company does have such a system then, for the purpose of recruitment, all a Manager will need to do is to use an existing description to derive a Job Definition. If Job Descriptions do not exist or if the vacancy is for an entirely new job, then one will have to be produced as a first step in the recruitment process.

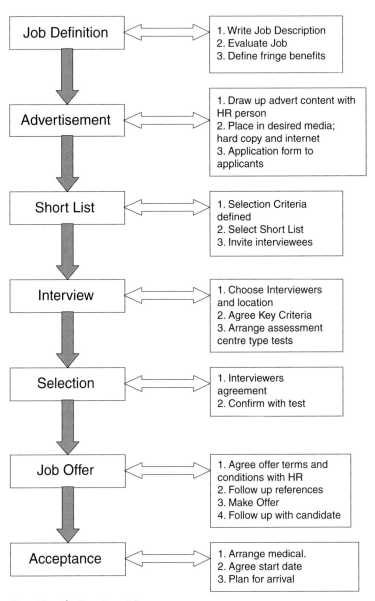

Figure A1. The Recruitment Process

Systems for the preparation of *Job Descriptions* vary from company to company but they all share common core features. These essential core features, described below, are the ones which will also be used as the basis for each step on the way to making a final job offer to the candidate.

- *Job Title and Purpose.* This must be precise and self explanatory, it is going to be used in the job advertisement, to attract only suitably qualified applicants, e.g. Synthetic Chemist to design and synthesise novel, organic materials for test as anti viral agents. It is also important to note where the job will be done, e.g. in a research group, in a plant team or as part of the technical service function etc. Similarly, it should be clear to whom the person will report, e.g. to the Team Leader.
- *Size or Dimension of the Job.* Includes both the financial and non financial aspects of the job, e.g., the job is part of a team, the holder will supervise two of the technicians within the team; financial responsibility is restricted to the purchase of consumables and monitoring against budget.
- *Key Accountabilities Features or Results.* These are the key areas where the job holder will need to achieve results, in order to fulfil the purpose of the job as given in the Job Title and Purpose, e.g. to develop laboratory processes suitable for semi technical manufacture in order to produce bulk samples of material for customer trials.
- *Special Features.* These are the qualifications and experience which are required to do the job, e.g. a graduate with three years experience; interpersonal qualities such as tact and diplomacy, persuasiveness, good communication skills etc.

If the job is entirely new then a Job Evaluation needs to be carried out so that, if required, a specific salary range can be given in the advertisement, and as a basis for the subsequent offer to, or in the negotiation of a package with the successful candidate.

The particular job evaluation procedure that is adopted will depend on the company, but all are designed to collect comparative data on jobs that are within a particular job market, e.g. UK chemical or pharmaceutical industry.

The procedures for job evaluation fall into two types [A-4, A-6].

1. Non-analytical methods, which assess each job as a whole.
2. Analytical methods, which score each job against a number of factors.

In the non-analytical job evaluation procedures the following methodologies are used.

- *Whole-job ranking.* Comparing the job with another and then creating a ranking order.
- *Paired comparisons.* Comparing a job with every other and assessing whether bigger, smaller or the same.
- *Job classification.* Each job is categorised against a standard job description.
- *Market comparisons.* Jobs are matched against standard job descriptions where data from the market is available.

Analytical procedures involve analysing the job against key factors and competencies and then scoring these using a standardised points system. An example of such a system that has been used commonly in industry for many years is the Hay method, and this is described briefly below.

The Hay Evaluation Method
The Hay Evaluation Method is based on the consideration that each function in an organisation is there in order to achieve results.

The following elements basically characterize any function:
• Know-how – "the sum total of every kind of knowledge and skill"
• Problem solving – "the amount and nature of the thinking required on the job"
• Accountability – "the answerability for action and its consequences"
• Working conditions (when appropriate)

The analysis of the function requirements makes it possible to do the following:
• Prioritise functions
• Qualify function contents
• Identify demand profiles for qualitative personnel planning
• Assess organizational structures, in particular gaps between levels
• Conduct objective salary comparisons

The possible applications of the method, especially in the field of staff functions outside the agreed scale or those in top positions, go considerably beyond conventional job evaluation procedures.
Evaluation Criteria: The method is designed for the analytical evaluation of functions by reference to a total of eight criteria, which comprise together three main groups as follows:

1. *Know-how*
    • Expertise or technical know-how
    • Management breadth
    • Human relation skills
2. *Problem solving*
    • Freedom to think
    • Complexity
3. Accountability
    • Freedom to act
    • Impact range
    • Type of impact

For each main criterion there is a two- or three-dimensional evaluation table. Evaluation is a process of allocating a function, in line with the demands of the function, to the definitions in the evaluation tables that best and more clearly apply to the function. Once the appropriate matching of the particular to the various individual criteria has been found, the relevant rating is derived as the sum of the following:

Know-how	X points
Problem solving	Y points
Accountability	Z points
Total Rating	X + Y + Z points

Profile
A comparison of the points scored for Problem Solving with Accountability determines whether the job is *intellectual-process* oriented or *accountability* oriented.

In reality the actual salary level for a particular job often depends on its importance
to the core activity of a company and other competitors operating in that specific sector of industry. For instance if a company needed to recruit the best scientists that
are available, it would place these higher on the salary scale than say mechanical engineers.

1.2.2
The Advertisement

The production, style and placement of the advertisement for the job involve a specialised skill. It is a very expensive activity and in larger companies will probably be
handled by the HR manager together with an advertising agency or even by an outside recruitment consultancy. Whichever is the case, it is imperative a Manager makes
sure that what appears in the text of the advertisement describes as precisely as possible the sort of person that is required and the job which is to be done. Later on there
will be the task of sorting through the application papers. This is arduous enough
without the added complication of attracting applications from inappropriately qualified people.
  Job advertisements are usually structured to cover seven essential elements.

1. *Image* – of the company
2. *Job* – what the job entails
3. *Need* – why the company needs to recruit
4. *Opportunity* – career opportunity for the applicant
5. *Profile* – what is required in the applicant
6. *Reward* – salaries and other benefits
7. *Application* – how to apply

Projecting the right image of the company helps attract suitable candidates and this
is often assisted by the style of the advert; its size, colour and graphics and whether
it is sober or jazzy. This is why this aspect is best left to the professional graphics
designers. Simply giving a job title in the advertisement is insufficient; and it must
say briefly what the job holder will actually do, as outlined in the job definition described in Section A, 1.2.1. Why the company needs to recruit is important since it
indicates whether the company is expanding and creating new jobs or roles within
the organisation. The geographical location must also be stressed as this has an important bearing on attracting candidates. Opportunities for career development are
the concern of most applicants for professional jobs. The opportunity should be outlined in a realistic and unambiguous way; it is no good saying "the sky is the limit".
Without providing a profile of the person likely to fill the post you will get quite un-
suitable people applying for the job. It has become very common to see comments
such as " salary and benefits commensurate with the post" or even more obscurely

"attractive packages". Whilst in advertisements from major companies this is satisfactory, since a candidate would expect first class remuneration and benefits from these organisations, in most other cases it is much better to state a salary range provided this is not too wide to be meaningless. Finally, the advertisement should contain clear instructions on how to apply; is it by an application form, in writing, by telephone or by e-mail to a named person; will a separate CV and names of referees be required etc.

Two typical examples of job advertisements for professionally qualified personnel that have appeared recently in the Royal Society of Chemistry's journal, Chemistry World, are given below.

> **Fuel our future**
> bp petroleum
> Based Pangbourne, Surrey
> With some global travel
> We are aiming to create more energy for the world without compromising the resources of the planet. Our global technology team is developing future differentiated fuels. Join us and you'll lead a range of technically challenging programmes and contribute to the development of fuels that have a reduced impact on the environment.
> **Senior Product Development Technologist**
> Leading programmes to support fuel developments, you'll use combustion fundamentals, combustion diagnostics, flow visualization and modelling toolsets, and deliver technical data based on fuel test programmes. You'll need at least five years' mechanical or combustion chemistry experience and a relevant degree, while a PhD in combustion chemistry or similar would be useful.
> To find out more and apply for this position, please visit
> www.bp.com/careers/ukexperienced
> Closing date: 14 March 2006

The company's desired image, that of a very large global and innovative energy company, is set out clearly at the beginning of this advertisement, highlighted by the "fuel our future" strap line. It is clear what the job entails and its geographical and organisational location but career development pathways are not indicated. The profile indicates what experience and qualifications are required and associated technical competencies are listed. Whilst the method of application is spelled out one obvious omission is any indication of the rewards to be expected by a successful candidate. One can only-assume that the absence of even bland generalisations in this area is because the drafters of the advertisement thought that it was not necessary in this instance because the job is with such a well-known, world-class petroleum company. The website referred to gives more details of the job requirement and in fact looking at this is essential for any applicant.

> **British Energy**
> Torness Power Station, East Lothian
> £ 21,999–£ 48,107 dependent on experience and qualifications

Assistant Chemist

British Energy is the largest generator of electricity in the UK, supplying energy to industrial and commercial clients. Employing more than 5000 people, our core business is nuclear generation. We own eight nuclear power stations in the UK with a combined capacity of 9600 MW and a 2000MW coal fired plant in Yorkshire. These stations supply 20% of UK electricity.

The work of the Chemistry team is varied and you will be expected to contribute to a wide range of task areas, it is anticipated that you will have a nominated role in the station emergency plan.

Major tasks include the following:

- Specifying chemistry and monitoring standards for the reactor gas and secondary circuits, together with auxiliary plant systems.
- Providing advice on matters of chemical safety
- Fulfilling the duties of the CoSHH nominated officer
- Preparing and reviewing technical standards, work instructions and other documentation for the chemistry programme.
- Assisting in the development of and delivering training in chemistry control and chemical safety

You should be able to demonstrate the following qualities:

- Excellent interpersonal and team working skills.
- A flexible and pro-active approach to work.
- A positive attitude to supporting continuous improvement.
- A high level of self-motivation, drive and proven ability in communication, influencing, motivating others and a "can do" attitude that focuses on safe and reliable operation.

You will be qualified to degree level or equivalent in Chemistry and have a thorough knowledge of chemical monitoring standards, techniques and application associated with nuclear plant generation. Preference will be given to candidates who can demonstrate a good understanding of AGR chemistry.

British Energy do not accept CVs. Application forms should be obtained from the British Energy website www.british-energy.com or by phoning the Resourcing Team on 01355 262727

This second advertisement portrays the company image as big in a big and important field of energy. The job is very specific and detailed. Similarly the profile for the applicant is very specific on experience and competencies. The financial reward is very broad and therefore runs the risk of getting applicants with varying degrees of experience within the job profile. At least this company is not afraid to put a figure on the salary, which is becoming increasingly uncommon, as opposed to the meaningless "competitive package" statement, which is now the accepted norm in such scientific jobs. Clearly this company do not want to attract a long list of applicants, they want a person who can fill the post with an absolute minimum of introductory training. The weakest point of this advertisement is the lack of any career development opportunities. One interesting point is the statement that CVs are not accepted.

A third example takes a very different tack to the two above; this one is from a major pharmaceutical company

Synthetic organic chemists

Merck Sharp & Dohme Ltd., part of Merck & Co., Inc., Whitehouse Station, NJ, USA is one of the world's leading pharmaceutical companies. Merck discovers, manufactures, and markets a broad range of products to improve human health, and has an unparalleled reputation for innovative research.

Our Development Laboratories, Part of our worldwide R&D function, are now looking for Synthetic Organic Chemists to join our Process research Group in Hoddleston, Hertfordshire. This department is responsible for devising new synthetic routes to pharmacologically active compounds and the supply of bulk drug for clinical trials and safety assessment studies.

PhD Chemists in Process Research

With a PhD and relevant post-doctoral experience, ideal candidates for these positions will have excellent practical skills and a broad-based knowledge of synthetic organic chemistry. Creative ability in planning and executing assignments in a multi-disciplinary environment is essential. An ability to work in teams as well as good written and oral communication skills are also important attributes.

You will have an excellent opportunity to broaden your knowledge of process chemistry and develop your technical skills in a stimulating collaboration with other scientists covering all aspects of the drug development process. MSD has a strong commitment to ongoing training and development. We provide first-class facilities and encourage our staff to maximise their potential through published work and conference participation. We also offer an attractive salary and a first rate benefits package.

Please write to Human Resources Department etc. or by e-mail etc with full CV.

This advertisement contains many elements that would make the position attractive to those with the desire to see their industrial work as an extension of their academic research. Emphasis is on their continued professional development, encouraging publication of their work and attendance at conferences, where such work could be further discussed or presented as a paper. This fits in with a long-standing tradition within the larger pharmaceutical companies, where the sharing of fundamental findings is considered to be beneficial all round.

In summary all good adverts should have a "hook", the *Opportunity*, with which to catch the quality person, only the latter of these adverts was specific in this respect; the others rely on the idea that working for a large, innovative company is a sufficient incentive. Of course, another highly important hook is the terms and conditions of employment, as exemplified by even the less than specific ones in the second case. However, the actual terms an conditions are often not within an individual manager's discretion being governed by company policy with regard to such things as the assessed value of the job, the position on the pay scale, having considered the experience of the potential recruit and, in extremis, how desperately the post needs to filled. With new graduates, there is often a variation in starting salaries offered by companies. This is based on the supply and demand in a particular discipline and in phase with the prevailing economic cycle.

Where the advertisement will be placed depends on what is appropriate in the geographical location of the company. There are basically six options.

1. The national press
2. The local press
3. General scientific magazines
4. House journals of the learned societies
5. Trade journals
6. Internet sites

The effectiveness of advertising in the national press will depend on the country you are operating within. It is generally better to reserve this for the more senior jobs. The local press is just the opposite; this is better for the recruitment of junior technical support staff, especially school leavers. General scientific magazines, e.g. New Scientist, are very good especially for the more rare disciplines, which are otherwise difficult to target. The magazines of learned societies are generally the best way of recruiting for a specific discipline, e.g. in the UK Chemistry World is held in high regard by recruiters of chemists. Trade journals, because they attract a specific readership, are often the way to attract experienced specialists for an urgently desired competency. Increasingly the Internet is being used to provide lists of jobs that can be readily scanned and applied for on line or by e-mail, and a website is now often the first port of call for people looking for new jobs. Many professional Societies now have lists of jobs on their websites available to members and non-members alike [A-19]. Quite often the sites contain just hard copy images cut and pasted into web pages but the medium offers a different way of attracting candidates and getting the message across using sophisticated graphics, video clips and links to related information such as company websites. The best of these are portals into company websites, where application forms can be filled in online and sent promptly to the correct person. However, this is not the place for the amateur and once again the best companies will have designer at hand to make sure the system works effectively.

1.2.3
Drawing up the Short List

Interviewing a large number of candidates on any reasonable time frame whilst continuing to do justice to all the candidates is impossible. Drawing up a short list of candidates from the many applicants produced by the advertisement of the job is essential.

When faced with a large number of application forms and letters, together with sizeable CV's, the production of a short list can be a daunting task. Under these conditions, dominated as it is by paper work, it easy to be influenced by the style and layout; "if the applicant can't produce a decent application form why should I bother to read it?" This is quite reasonable response but Managers should try not to be too put off by a rather poorly produced application, but concentrate on the more important aspect, the content. Candidates should be encouraged to produce succinct and readable applications, and a well-designed application form can help considerably in this area. Whilst application forms are no substitute for a well written CV, it should be recognised that the latter are a form of self-advertising and an application form can seek only the pertinent data, in the form of a résumé CV, for quick scanning. Application forms can be made available to candidates in down loadable form from web sites or be delivered by e-mail, enabling completion in a word processing package. Specifying the type of font and size on these forms avoids receipt of undecipherable hand written and overlong letters of application. Increasingly companies are specifying that application forms be filled in on-line company websites.

Rejection of potentially good candidates can also be avoided if there is a sound procedure in place, preferably one that has stood up well on previous occasions. The work that has already been done at the Job Definition stage, where the critical or key criteria were identified, will be of great help in the initial screening of the applicants.

Applicants who are missing any of the key criteria for the job should be rejected. For instance, if the object is to recruit a person who can immediately supply a particular skill or experience, applicants who do not have these skills should be passed over, even if on other counts they look attractive. If there is time to train an individual for a particular job, or a person with management potential in the medium term is being sought, these other factors, such as interpersonal skills, will be more influential when selecting for the short list.

With graduate recruits, the nature and class of their degree will be an influencing factor. Additionally, the institution where the degree was obtained is likely to be relevant. For postgraduates, higher relevance is usually placed on the research aspects; the subjects on which they have carried out their research, who was the academic supervisor for this research or where any postdoctoral study was carried out. These factors may well determine whether they are selected for the short list.

If you there are any doubts, it is useful to talk to applicants on the telephone or even to meet them on neutral territory for a brief discussion. At the end of the day Managers need to be sure that they have selected the best people for the short list. Remember, it is often said, *"past behaviour is the best indicator for future performance"*.

What then is the most appropriate size of a short list? Well four is a very good number but it is usually considered to be reasonable to interview between four to six people in one day. If all the preparation work has been done well and there are a good number of applicants, this number should be enough to satisfy a single vacancy. When recruiting several people for similar jobs, it is sensible to have some multiple of the above figures, and to spread the interviewing over more than one day.

1.2.4
Selection Interviewing

Selection interviewing can be very stressful, for both interviewer and interviewee, and every attempt should be made to minimise any detrimental impact of this stress. Whilst technology is readily available for interviews to be carried out via some form of video conferencing, in most companies face to face interviewing is still the preferred method.

Interviewers should be happy that all the required paper work on the candidates, e.g. application form, CV etc, is available; that all the necessary physical arrangements, e.g. transportation and accommodation for candidates have been made and that the rooms for carrying out the interviewing are available.

Interviewees should be made to feel comfortable with the process and one simple and effective way of doing this is to meet all the candidates in a relaxed social context prior to the interviews. For instance, with professional candidates, this could involve having a meal together the evening before the interviewing day. Including in the party a member of the potential peer group from the laboratory or similar working environment helps to relieve the tension amongst the candidates. These social gatherings are a good way of picking up any behavioural aspects, e.g. verbosity, shyness which may not be apparent or, conversely, greatly emphasised in the more formal atmosphere of the interviewing room. Managers can use these observations to adapt or modify their interviewing styles to suit each candidate and get the best out of them during the interviews. It is only the style of the interview that is modified, not the process of interviewing, which should be consistent for each candidate in order to make the post interview comparisons valid. Time is also saved by using these social occasions to introduce the company, its structure and format, and in answering any general questions that will help the candidates to understand the role of the R&D group within the company or its particular business function.

The formation and operation of the interviewing team is the next topic to be resolved. Some organisations prefer the interview to be carried out by a panel of people with the candidates answering questions put to them by this panel in a single session. It is this author's view, that for R&D personnel these occasions can be intimidating and it is much better to carry out one to one interviews. An exception is if a candidate is required to give a technical presentation. This should be done in front of a group of suitably qualified personnel; a mini seminar with the opportunity for probing questions regarding technical competence.

The ideal membership of the interviewing team should be the R&D line manager, a Human Resource manager or Personnel Officer and a person who is technically expert in the work area where the recruit will be employed. In larger R&D Groups it is often better to split into two teams; the managerial people looking at the behavioural and interpersonal aspects and a technical group, as already stated, looking specifically at the scientific competence of the candidates.

Having selected the members of the team a Manager must ensure that they all fully understand what is the purpose of the job and what it entails. Having a pre interview briefing session does this. During this briefing the technical and non-technical skills that are required in the potential recruit must be clarified. To help in the process of defining the non-technical criteria, many companies use packages put together by management consultants. However, if the company does not have such a package available it is relatively straightforward to draw up a list of the job requirements.

Typically within R&D these will be:

- Innovation
- Creativity and lateral thinking
- Analysis
- Job motivation, particularly self motivation and initiative
- The ability to plan and organise
- Communication skills, both written and oral
- The desire to carry on learning and self-educating

For a managerial position, or a person with management potential, the additional requirements will be:

- Leadership, especially team leadership
- Self control and sensitivity to others
- Judgement and the ability to make decisions
- Tolerance to stress and working under pressure
- Energy and skill to develop members of the team

To illustrate the process that can be adopted and the allocation of data within the interviewing team let us consider the following example.

> A pharmaceutical company requires a postgraduate chemist in order to boost the level of synthetic chemistry in an R&D team working on drugs for use in a particular therapeutic area. After a few years in this role the person will be expected to become a team leader and then possibly move into R&D line management.
>
> - The technical team are looking to see if the candidate has a high level knowledge of chemistry and synthetic methodology, an understanding of medicinal chemistry and biochemistry and possibly the particular therapeutic area.
> - The management/human resource team have agreed the key criteria and they are using these in the form of a checklist for their questions to each candidate. In this case they have chosen to home in on oral and written communication, team leadership, job motivation, judgement and decision making together with planning and organising. They have asked the technical team to also consider innovation, creativity and analytical skills.

When carrying out the individual interviews the questioning must lead to meaningful answers from the candidates; must be open ended and require much more than yes and no answers; must avoid leading to a particular answer. Vague generalisations are also of no use at all. Candidates should be asked probing questions and required to provide examples, which will illustrate their answers including the outcome, results or effects of any actions. It is only by getting this type of response that the interviewer should be convinced that the candidate really has the experience needed to perform the job. At the end of each interview, positive examples against each of the criteria allocated to each interviewer should have been obtained. This data will be used in the post interview discussions that will be held by the other members of the interviewing team.

1.2.5
Psychometric Tests and Assessment Centres

Support for the evaluation of the non-technical make up of a candidate is very commonly sought by carrying out some form of *psychometric tests* on the candidates. Psychometric tests are designed to assess either ability (cognitive tests) or personality characteristics and there are three basic types: ones to assess a specific ability, e.g. numeracy, verbal and spatial perception; ones to test general mental ability, e.g. analytical reasoning, creative thinking and those to asses general personality traits, e.g. learning styles or team types [A-4, A-7]. These tests are usually done by a Personnel Officer who will have been trained to do this, or alternatively, but more expensively, by employing an outside consultant (see also Section C, 1.5.1).

Whilst it is relatively easy to carry out, the correlation between the test results obtained on a candidate and the actual performance on the job, known as *predictive validity*, is variable and depends on what characteristics are being looked at in the test. The correlation coefficient for cognitive tests has been assessed as 0.35 and for personality tests as low as 0.15 [A-4]. However, when such tests are well done and analysed by a suitably qualified person they are very useful in checking or supporting the conclusions of the interviewing team, particularly where there is a divergence of opinion, but they should never be used as the sole basis for selecting a candidate.

Because no single test is sufficient in its own right for making a judgment on a particular candidate it is better to use a combination of different techniques. This combination of techniques is known as an *Assessment Centre*.

An assessment centre is not a physical place but is a suite of exercises designed to assess a set of personal characteristics. Their use dates back to the Second World War when the British War Office developed them to select people with officer potential from within the ranks. Nowadays they are used quite commonly in medium to large size companies not only for job selection, specifically graduate selection or managerial appointments, but also increasingly in internal promotion and development of staff for managerial careers, when they are more often called *development centres* [A-8].

A range of tests are used in an assessment centre.

- Structured interviews
- Psychometric tests
- In-tray exercises
- Group discussions
- Group problem solving job simulations
- Individual job simulated tasks
- Job related role play

Typically for graduate recruitment the assessment centre will be run over two days and involves a high usage of management time. This is because, unlike psychometric testing, many of the exercises cannot be scored by standard methods and involve observation of the candidates performing certain roles and carrying out specific tasks by a team of assessors. This team usually comprises line managers and personnel officers, trained in the methodology, and perhaps an external training/occupation psychologist acting as a consultant. It is very important that the tests chosen are those needed to evaluate the required competencies for the job; those identified in the job definition stage and selection criteria (see Section A, 1.2.1 and 1.2.5). Once chosen these skills and competencies are rated in importance and scored or valued accordingly and a matrix drawn up for use by the assessors. During the exercises the assessors can observe the behaviour of the participants and then score them against the matrix of the key competencies [A-9]. At the end of the exercises the assessors can compare their scores and the results moderated for each candidate.

1.2.6
The Choice

Whatever interviewing and/or assessment methodology is used the crucial point arrives when the decision on who to appoint has to be made. When using an interviewing team it is very important to come to a consensus view and not to let the views of a dominant personality hold sway. Consequently, the views of each member of the team should be canvassed and taken into account when discussing each of the candidates. In selecting for R&D jobs there is one exception and that is over the matter of the scientific and technical competence of the candidates. If on consideration the technical team have expressed doubts on the current technical competence or the ability to respond to the required training, or even more importantly the inherent scientific ability of any of the candidates, then they should be rejected in the first screen. Only when these competencies for each candidate have been agreed as being satisfactory should the other aspects required for the job be evaluated. This evaluation involves assessing each candidate against the list of criteria set before the interviews, which were based on the job description. Confirmatory information is obtained from psychometric testing or from full-blown assessment centre results.

Returning to our example of a chemist required by a pharmaceutical company. Four candidates were eventually interviewed.

- The technical assessment team had found that Candidate A did not meet the standards required for the job. The person was very narrow in approach to chemistry knowing a great deal about the PhD topic but little else, even some basic tenets were poorly understood. Additionally little enthusiasm had been shown for medicinal chemistry. Candidate B had given a rather poor presentation of current research work but under general questioning was found to have a good knowledge of chemistry and a keen analytical mind. The remaining two, Candidates C and D both displayed high levels of competence and skills.
- The managerial assessment team considered that Candidate A was a good communicator having edited the student magazine, including leading a team of reporters, could make decisions and was good at planning and organising. Candidate B was indeed a poor communicator and was really a team member rather than a potential leader. Candidate C was well organised, could plan and make decisions and could act as a team leader having demonstrated this by organising a group of students to do social work in the local community. However, they had some doubts about the motivation of C to continue in a career in chemistry. Candidate D was found to communicate poorly when dealing with topics other than chemistry, but showed high motivation to continue in research. Candidate D was also good at planning and organising but had not yet had to make any hard decisions and so far had not demonstrated any team leadership skills.
- During the post interview discussion the management team agreed to drop Candidate A on the recommendation of the technical team, even though they

considered that the person had management potential. Candidate B did not have the potential to become a team leader and was indeed a poor communicator. The choice was therefore between C and D. On the basis of the psychometric tests it was concluded that candidate D was not a team player but a person who liked to act as an individual and was very single minded, potentially a good researcher but not a manager. The only worry about Candidate C was the long-term commitment to a career in chemistry. The psychometric tests indicated a person who would be very loyal when once accepted in to a group.

- The decision was made to offer the job to Candidate C.

Any lingering doubts or questions over the chosen candidate can always be taken up by arranging another brief interview before making the final offer. This is done preferably in person but a telephone conversation is acceptable if there is some urgency. It is also better, if there are any lingering doubts or uncertainties, to sleep on any decision and revisit the data the following morning, in the clear light of day. Using this method the position is avoided where everybody, after all the hard work, is keen to see a successful conclusion to the interviewing process, but in reality none of the candidates are really up to the mark.

1.2.7
The Offer

If a choice can be made on the day, then it is simply a matter of informing each candidate of the decision. However, if interviewing of the candidates occurs at different times or over several days then this is not possible. In this situation, having confirmed with the HR manager the terms and conditions that are going to be put in the offer letter, it is good practice for the Manager to make a personal phone call to the preferred candidate. This serves several purposes. Firstly, it checks whether the person is likely to accept the offer before any of the alternative candidates are rejected. Secondly, if there are any worries or queries on the candidate's part, they can be answered promptly, before they loom to large in the candidates mind. Thirdly, if the candidate wishes to have more time to consider the position, a time limit can be agreed by which a firm answer must be received. It is also necessary at this stage to remind the person that the offer will be *"subject to obtaining satisfactory comments from their nominated referees"* and also a medical examination. They can then be told that the formal letter from Human Resource department will be sent to them very shortly.

Written statements of terms and conditions are required by the 1998 Employment Rights Act in the UK and similar legal requirements exist in other countries. However, under UK common law a contract exists when two parties reach an agreement which both intend to be legally binding or when the contract is supported by something of value passing between the parties. In the employment area the something of value would be the salary. Neither case needs the existence of a contract; oral agreement suffices, but without a written contract arguments could arise in the future. At the very least the formal job offer letters should describe the terms and conditions and require employees to sign one copy and return, keeping a second for their own records [A-4].

Why have the referees not been consulted before this point in the process? Simply to avoid causing any embarrassment to the candidate. Problems could arise if, for instance, one or more of the referees were from the candidate's current place of employment. There is little point in causing any problems for somebody who is not going to be offered the job. In the case of postgraduates looking for their first job, it is permissible to consult with their academic supervisors at an earlier stage, especially if a Manager is on good terms with these people and values their opinions. It is rare that a person claims to have qualifications that are bogus, but if you want to be entirely confident it is possible to get written confirmation from the relevant professional body, institution or university.

All the unsuccessful candidates should be sent rejection letters promptly. It is also courteous for a Manager to make a telephone call to these people and be prepared to answer any queries they may have about the recruitment process. It is also fair that their reasonable requests to provide an explanation of why they failed to get the job are met. This is a matter of good manners and requires little effort on a Manager's part, usually just a few minutes on the telephone. To do this some decent notes from the interviews should be retained to use when these requests arise. The only thing that must be avoided, and must never be allowed to happen, is any questioning of the decision on who has been appointed.

1.2.8
Acceptance and Induction

The arrival of the acceptance letter is the high point of the recruitment process. It is the successful conclusion to what is a long and difficult process for all concerned. However, it is not the end of a Manager's activities. A plan should be made for the arrival of the new recruit.

Practical issues need to be resolved, such as office and laboratory space to be arranged. First day induction activities, including safety training, company and legal health and safety policies for those working in the laboratory, being given a guided tour of the site, time recording policy and whatever else is in the company's preliminary programme for new recruits must also be organised. Induction is not done solely because it is a legal and ethical responsibility of management but also to make the recruit feel welcome and valued by the organisation.

Detailed discussions between the Manager with the person who is going to be the recruits immediate supervisor or mentor should be carried out and an initial work plan agreed. Outline details of this work plan may then be communicated to the recruit, especially a postgraduate, who will then be in a position to make a confident start in the new position. A note is put in the diary of the start date so that the Manager is available to welcome the new addition into the company and to address any immediate concerns. Introductions to members of the group in which the new recruit will be working should be made personally.

2
Developing the People who form the Skills Base

Managing an individual's performance to meet the desired objectives is a necessary step.

The Skills Base in any R&D organisation is made up from the talents of the people who work within the group, i.e. it is a summation of their individual competencies. In the first part of Section A the importance of matching the R&D Skills Base to the needs of the business was described. Having defined this Skills Base, it is the task of the R&D Manager to ensure that the scientists and technicians, who are its constituent parts, continue to develop, so that they are able to perform at the level required to meet the group's objectives.

There are three main thrusts of this management activity. The first is the setting of measurable work objectives, commonly known as *performance criteria*, for the individuals, and the provision of the necessary training or coaching to enable that person to fulfil such criteria and achieve the desired targets. The second thrust is the managing of the individual's performance, *performance management,* by regular appraisals and reviews. The third thrust is that of *career development so* that personnel can achieve their potential wherever possible. This latter point is obviously of great importance to the individual but it is likewise for the organisation. Every organisation matures and changes, and a *succession plan,* which is based on the career development of its staff, is much better than change created by a series of emergency appointments.

2.1
The Management of Performance

In the last decade of the 20th century the reward systems operated within most companies changed from those involving an automatic rise within a particular grade until a maximum is achieved, to those that are more directly linked to the performance of the individual in the job. This reward method is known as *performance related pay* (PRP), which can take several forms. In many companies the grade for the job still applies but there is a much wider individual merit band, ranging from 80–120% of the job grade. An individual can move up and down within this band, depending on their performance appraisal as part of an annual salary review. In others, bonuses are

Research and Development in the Chemical and Pharmaceutical Industry, Third Edition. Peter Bamfield
Copyright © 2006 WILEY-VCH Verlag GmbH & Co. KGaA, Weinheim
ISBN: 3-527-31775-9

paid to both individuals and teams based on the actual performance of the company or a certain business sector in a particular year. Whatever the reward mechanism, for PRP systems to work in a fair and equitable manner, it is necessary to have a personnel process that includes sound methods for evaluating an individual's performance, and hence their contribution to a group's results. These personnel processes are known as *performance management systems* [A-10].

All employees do not universally like performance related pay. Cogent criticisms have been made by some independent commentators against many aspects of the way in which performance related pay has been introduced into some companies. These criticisms include, the setting of unrealistic performance objectives, inconsistent appraisals, apparent favouritism and inadequate funding [A-11]. A line manager within the company should recognise that these criticisms, often based on peoples perception of the system, exist and must work to minimise their impact. Any hint of unrealistic performance criteria being set, that the management line is being inconsistent in carrying out the performance appraisals, and there will be problems.

Some scientists working in R&D often say, or at least think, that there is a particular difficulty in judging their performance. They consider that the type of jobs which they perform, ranging from the longer term, strategic or so called 'blue sky' research to the very specific provision of a service such as mass spectrometry, are not appropriately handled by a general company system. It is true, that in the more service orientated jobs it is relatively easy to put measurable parameters on the output, in a similar way to a commercial or manufacturing post. For people in the more speculative research jobs this is very much more difficult but, with some minor adjustments, PRP is eminently suited to R&D work. The selection of the performance criteria and the methods of measurement to be used for R&D jobs do need to be carefully thought about. The point to be remembered is that these criteria must be just as achievable as those set for any other types of job. R&D Managers must recognise that, for those people at the more speculative end of the R&D spectrum, the difficulty in providing meaningful quantification to support their performance does lead to some worries, as emphasised elsewhere in this book. A fair and effective setting of performance criteria, which they can both understand, agree with and be able to self-monitor, can allay these fears. The performance criteria for support technicians, for instance people who are carrying out day to day practical work, will also be very different to those set for the professionally qualified member of the group. The same management standards must always apply when setting the criteria and judging the performance any group of staff.

The basic method which is used in performance management is as follows (see Figure A2):

- Each member of staff has an individual job description that includes accountabilities, objectives and key result areas (see Section A, 1.2.1).
- Based on these job accountabilities and key result areas, personal performance expectations or standards, including a time frame for their achievement, are set in agreement with the job holder.
- Training, which may be required by the job holder in order to be able to achieve the designated goals, is discussed and then provided.

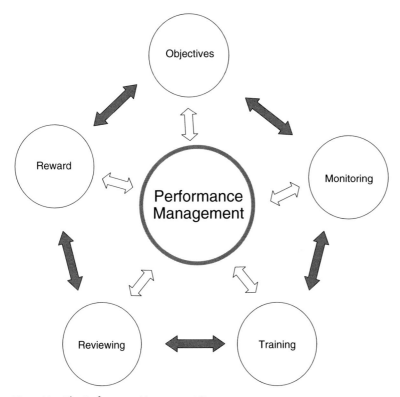

Figure A2. The Performance Management Process

- Performance against the targets is then reviewed regularly with the job holder (*performance appraisal*), using data collected during the intervening time period, and the targets modified as necessary.
- The results from the reviews are then used in determining the reward for the individual and in determining career development opportunities.

Even if the company does not operate a full performance related pay system, R&D Managers can still apply the principles used in performance management to good effect and hence get the best out of their staff. However, to be really effective, performance management must in some way be connected, however loosely, to a system of reward.

2.1.1
Setting Performance Objectives

The basic task for the R&D Manager is to make sure that performance objectives and criteria for success, whilst being sufficiently difficult to stretch the individual, are achievable within the desired time frame. There is no point in setting performance criteria that are so easily achieved as to be meaningless, nor so difficult as to be unreasonable. These criteria must be agreed with the individual concerned, and are best set by the person's immediate supervisor. Problems will arise if supervisors are not working to the same standards and format for the process, and this aspect will need to be monitored. Performance objectives and criteria for success should conform to the following.

- Stretching for the individual, but not so difficult as to be unreasonable
- Achievable within the desired time frame
- Agreed with the individual concerned
- Set by the person's immediate supervisor
- Based on common standards across a group
- Clear who will act as a monitor/mentor
- Once agreed are used to monitor performance over the desired period

A training package for the performance management system adopted by the company will no doubt be available in larger companies. The Manager should make sure that everybody has been through the appropriate aspects of the training package, before they get involved in the setting of performance objectives. This does mean everybody. The assessors and the assessed must have undergone the necessary training so that they both speak the same language. The very minimum training should be to ensure all staff is working to a common standard of assessment.

For the purpose of exemplifying the method involved, a vertical slice through the organisation in the process development department of a fine chemicals manufacturing company will be taken. The actual process is very detailed and for reasons of brevity, only an outline is given here rather than the full details.

The line manager relationship is as shown below:

Head of Department
Group Leader
Process Chemist
Laboratory Technician

Head of Department and the Group Leader
In order to set the Performance Objectives for the subsequent period the Head Of Department has arranged a meeting with the Group Leader. The HOD has the Group Leader's Job Description, containing the accountabilities and key results areas, and that part of the Departmental work plan which has been allocated to this particular Group.

The Group Leader manages three process chemists and one process engineer. The team also has five technicians as the support staff. The group is charged with the overall responsibility of converting laboratory processes into ones that can be operated on a full scale manufacturing plant, sometimes via the use of a pilot plant.

The HOD sets the following performance objectives:

Technical
1. Products A and B to be ready for manufacture in three to six months time respectively.
2. Pilot plant trial for Y to be complete in six weeks.

Personnel

1. Ensure the Performance Objectives are set for all staff within the Group in one months time.
2. Monitor and review the performance of the staff at bi-monthly intervals.
3. Determine the training needs of all staff and arrange for this to be provided throughout the year. Keep the Training Records up to date.
4. Make recommendations for individual rewards and career development at the Annual review in six months time.
5. Carry out formal safety audits in the laboratory every month.

Budget and Plans

1. Monitor expenditure against the Annual Plan. Advise on any potential over-spend each quarter.
2. Determine the Capital Expenditure requirements of the Group.

Personal

1. Improve negotiating skills e.g. with plant management.
2. Make effective presentations of the work of the Group to senior management, both written and orally.

Training

Course on Negotiating Skills. Coaching on presentation (HOD to arrange).

Group Leader and the Process Chemist

The Group Leader then carries out a similar exercise with one of the Process Chemists:

Technical

1. Develop a process suitable for the scale up manufacture of Product B in three months.
2. Work with the Chemical Engineer on the design of the pilot plant for Product B, to be completed in four months time.

Personnel

1. Set performance objectives for the Technician for the coming period.
2. Review and monitor the performance of the Technician monthly.
3. Provide coaching to the Technician on laboratory skills, safety and plant operations. This to be done on an ongoing basis.

Personal

Develop interpersonal skills and use these when dealing with plant management.

Training

Time Management course (Group Leader to organise)

Process Chemist and the Technician

The process Chemist then follows the same process with the Technician:

Technical

1. Carry out experiments as requested and enter details in the laboratory note-book daily.

2. Check and order chemicals and other supplies needed for the laboratory on a daily basis.
3. Clean and calibrate the chromatography equipment weekly; report faults promptly as they occur.

Personal
1. Improve written work.
2. Work more tidily in the laboratory.
3. Broaden personal knowledge of synthetic methods.

Training
1. Report writing (supervisor to arrange).
2. Synthetic methods (coaching by supervisor).
3. Use of automated synthesis equipment (supervisor & senior technician).

These performance objectives will then be reviewed at the agreed times, with each level of staff and modified, depending on the results, with new objectives added as appropriate. This is a continuous improvement cycle.

2.1.2.
Reviewing and Monitoring

Regular reviewing against performance expectations is something that will take up a considerable amount of an R&D Manager's time. It is an activity that is beneficial to all concerned. It provides a Manager with the opportunity to discuss, directly with an individual, any problems or difficulties that seem to be inhibiting the progress of their work. It is a time when training, coaching or other actions can be identified that will help to resolve these difficulties and improve the performance of the individual. This reviewing and monitoring, although a part of the formal system, should not be restricted to these pre-arranged occasions. Reviewing and monitoring should become a natural way of life and be part of the day-to-day contact between a Manager and the managed. With time, it is possible to develop ways of carrying out this process without it becoming too intrusive, which is essential. This regular contact may simply take the form of asking the question, "how are things going?" during chance meetings or spontaneous visits to the laboratory. It is important to ensure that the person feels involved, seeing these questions as representing a genuine interest on the Manager's part, not just a passing politeness. The individual will only raise real problems or difficulties, if the relationship is seen as a two way process. These informal talks also provide an excellent opportunity to give praise for a job being well done.

In R&D a Manager needs to take particular care and show genuine patience when dealing with technical achievement. The old maxim that "positive results from research are not guaranteed" should always be remembered and encouragement is the name of the game. If a chemist is making a strenuous effort to overcome a serious technical problem, the last thing that person wants to receive is a reminder that the result was promised for next week. These interpersonal skills are further discussed in Section A, 3.1.

Quantitative measures, or metrics, are surprisingly easy to produce for R&D work but they must be the right ones for the purpose. For instance, within targeted research the number of samples submitted for the primary test or screen and the number accepted for second phase testing is easily recorded. In reality these measures are typical of the ones that need to be used with extreme caution when judging performance. Simple quantity is inadequate; the real objective is to get to the lead candidate in as few experiments as possible. However, the target for any performance measurement is for it to be quantitative, wherever possible, but always remembering a quality qualifier is required.

A lot of the data collected on an individual's performance will be subjective. One important source of this type of information is from third parties, those people who come into contact with the person during their routine activities. For instance, it could be obtained by having discussions with plant managers, colleagues in marketing, consultants or academic collaborators. All these people can provide valuable, independent views on the performance of both individuals and groups, beyond the rather narrow technical confines of the R&D function (see 360° appraisal in Section A, 2.1.2.1).

At the time of the formal review with a member of staff, as much information as possible should be to hand. This will ensure that it is a positive discussion. Whilst spending some time discussing poor performance cannot and should not be avoided, the objective is to build on examples of good performance.

2.1.2.1 Performance Appraisal Schemes

Performance appraisal schemes fall into the four generic classes shown below.

1. Traditional appraisals
2. Competency based appraisals
3. 360° appraisals
4. Upwards appraisals

The traditional appraisal scheme involves the manager appraising the performance of staff members on a one to one basis. These can be well-conducted conversations on an individual's performance but do have the capacity to become authoritarian and dictatorial. Competency based appraisals are those where performance is appraised against clearly identified competencies and are not concerned with personality or character traits. 360° appraisal, as its name implies, involves getting feedback on a person's performance from contacts all on fronts. Upward appraisal is a method used by managers basically to improve their personal performance. Getting the views from their own staff on their managerial skills does this.

Each of these appraisal schemes has both benefits and drawbacks and these are summarised in Table A4 [A-10]. Which appraisal scheme that is used is usually decided at the highest level in the human resources department of the bigger companies. In small businesses the choice may will be up to the manager concerned; whether they wish to operate via traditional command and control hierarchy or whether they are customer focused with members of staff self empowered.

Table A4   Comparison of Performance Appraisal Schemes

Scheme	Positives	Negatives
Traditional	Performance overview achievable One to one review Records easily kept Action plans made Training needs identified Simple to operate	Feedback can be one way Authoritarian temptation for managers Daunting if staff differential too large
Competency Based	Competencies easily identified Consistency across organisation Common language/definitions used Specific self assessment Objective Action plan for training/development	High levels of management/staff input Coordination to be high between HR and Managers All staff need to be inducted Managers to be trained to same standard
360°	Holistic approach All round feedback Cultural change – allows rapid response in changing world Achieves business objectives	Complexity Most intensive in research and preparation Needs a no-blame culture May raise expectations that are difficult to deliver
Upwards	Two-way process Offers insight and awareness Identifies strengths and weaknesses Customer focused culture helped	Non supportive company ethos Time consuming and complex Danger of being used to settle old scores

For the increasingly popular 360° appraisal scheme the choice of people to provide the necessary all round feedback to R&D personnel is once more dependent on the role within the group [A-12]. Is the work carried within a multi-disciplinary team? Is other staff directly managed? Is there contact with external customers or suppliers? What contact is there with manufacturing or other functions? Is there collaboration with academic institutions? A typical range of possible contacts for feedback on the Group Leader (see Section A, 2.1.1) is illustrated in Figure A3. It should be noted that this contains an element of upward appraisal as the opinions of managed staff are also canvassed.

Having made the choice of opinion providers it is necessary to ensure that they understand the process and purpose of the exercise. Either the Manager or HR person needs to provide a clear explanation. To help the process it is common to use a standard form. In this way responses can be compared from the different sources. When completing the 360° forms the respondents should be asked to observe the following.

- Be objective
- Provide examples
- Be constructive in feedback

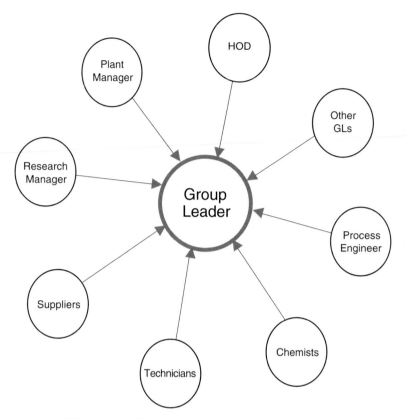

Figure A3. 360° Appraisal of Group Leader

- Non judgmental, avoiding negative criticism
- Try to be motivational

Providing the feedback can be quite testing for both the receiver and the giver of the information. The receiver being made to understand that the process is designed for performance improvement and career development, and will be linked to training where necessary, can ease this difficulty.

2.1.3
Training and Development

The modern chemical and allied industries are in a constant state of flux. Business portfolio changes occur more rapidly than in the past; new technologies arrive on the scene with increasing frequency, with the consequent danger of obsolescence for existing competencies and staff. In this environment managers and staff need to be continuously upgrading their skills by undergoing *continuing professional development*.

Whilst the continuous development of staff is very important within any department of a business, for R&D departments it is an absolute prerequisite in order to avoid any loss of competency. Today research staff also has an increased awareness of the need to get their innovations to the market as quickly as possible. To achieve this goal they need to be exposed to the world of business and commerce to a much greater degree than in the past. Consequently, a higher level of training in these areas is required so that they feel confident in their dealings with potential customers for their technology. It should be noted at this point that the provision and recording of all training is a formal requirement of any quality registration system. This aspect is covered in Section B.

As mentioned above, the training needs of each individual are identified during the performance review process or through assessment centre techniques. Essentially this is the training required by the individual for the purpose of career development. It is extra to that which may be provided in any corporate core-training package designed for all members of a particular group or level of staff in a company. The following questions need to be addressed in the review process.

- What information does staff need to have in order to perform their jobs to a high standard?
- What skills and competencies and to what level are required?
- What attitudinal characteristics are required?

Within R&D the answers to these questions fall into the following four main categories:

1. Technical Skills
2. Communication Skills
3. Interpersonal Skills
4. Safety and Quality

Technical skills are of prime importance to all personnel within R&D and therefore merit the highest attention by the Manager. The training requirements of the individual member of staff will depend on their level and role within the organisation

Younger technicians are very likely to be undergoing some form of vocational training. This could be as a part time student at a local college of technology or university, by distance learning, by a course within the company or by a correspondence course at home. Whichever is the case, this can be quite a difficult process of education for the individual. It often helps, if possible, to identify somebody who can act as a mentor for these people. The mentor, who is there to offer help and guidance to the trainees and provide a link into management, will usually be the qualified line supervisor. In multi-disciplinary teams this may not be possible and a sympathetic alternate may need to be found from another group. Apart from the formal academic education, the technician level staff will typically require skills training in the handling and maintenance of laboratory equipment, a variety of analytical methods and in aspects of IT and basic information retrieval. This type of training is best done on the job, as it is required, otherwise it will be quickly forgotten. Similarly, the tacit knowledge of senior technicians, their know-how, can only be passed on by specific mentoring on a one to one basis.

For the professionally qualified staff, technical training is more a question of on-going education and specialist courses. These are often provided by learned societies, such as the American Chemical Society, the Royal Society of Chemistry or the Institution of Chemical Engineers, in the form of workshop type meetings and symposia. With older scientists, refresher courses will often be required. Trade associations, such as the Paint Research Association, also provide specific technological courses. Additionally, there are very many commercial organisations offering a variety of courses. It is essential that scientists keep up to date, and also meet and discuss with their equivalents in other companies and countries. Attendance at relevant international symposia is a good way of satisfying this requirement. It is difficult to do an exact cost benefit analysis on these training activities, as their effect is cumulative and not immediately measurable. However, it is abundantly clear, if time is not allowed for this type of training, gradually, over a period of years, the capabilities of the scientific personnel base will be diminished.

Communication skills go beyond the requirement for scientists to be able to construct and produce a well thought out technical report. Most scientists are required to stand up in front of an audience and describe their work in a clear and unambiguous manner. Traditionally this has been to their peer group of scientists but increasingly they are required, as part of the business team, to talk in front of customers both at trade affairs and on technical visits. The ability to present their work to people who are not experts or even knowledgeable in their branch of science is also seen by management as a desirable skill. This latter communication skill is mostly used in dealings with local community organisations and environmental pressure groups.

The interpersonal and personal effectiveness skills relevant to R&D staff are basically a sub set of the general ones for all employees. Those chosen for development in a member of the R&D group will depend on the current role and level of that person in the organisation, and those that will be required to meet the individual's career plan. Typically, for professionally qualified staff, they would include:

- Effective negotiation
- Team working
- Supervisory management
- Planning and organising
- Project management
- Business awareness

Safety training is a legal requirement in all developed countries (see Section B). Therefore, all staff in the laboratory will have to undergo special training in safety, from the day they arrive as new recruits and on an ongoing basis. The R&D Manager will be responsible for ensuring that the company policy in matters of safety is applied to the letter in the laboratories and semi technical plant. There are no prizes for cutting corners on safety.

Total Quality Management (see Section B), especially opposite the basic requirements to meet the ISO standards such as ISO 9000 etc., is a similar area where the company policy must be followed and the necessary training be provided. A specific requirement of TQM will be the maintenance of accurate and up to date training records for all staff. The provision of a safe environment and Quality in R&D will be dealt with in greater detail in Section B, 3.2 and 3.4) on organisational issues.

2.1.4
Reward

It is impossible to go into detail about the rewards which should be made to people for their performance in the absence of the policy of a specific company. Therefore, comment at this stage will have to be restricted to the general principles behind any reward system.

When operating a performance related pay system, the rewards made to individual members of staff should reflect their personal performance, as discussed with them at review time. If this is not the case then over a period of time a Manager's judgement will be called into question. This could well lead to dissatisfaction and to a subtle state of non co-operation, even amongst what is normally a highly motivated group of people in R&D.

The value and nature of any non-monetary rewards will depend on the culture that exists in the company or country. However, it is universally accepted that a genuine word of praise is welcomed and valued by everybody. A very strong motivation for scientists is the recognition of their scientific work by their peer group. The R&D Manager should seek ways of demonstrating this at every opportunity. For instance, being on the lookout for prizes and awards made by professional bodies or government agencies etc., which may be relevant to the staff concerned. Even nominating a person or team of people for an award can give great satisfaction to the individuals involved. If a member of staff or a team obtains an award, it should be given maximum publicity to reinforce the message internally in the company.

There is an inherent problem for R&D in all performance rewards made to individuals. In many areas of R&D the work is increasingly being done by people working in teams, either as a sub group of the R&D group or cross functionally within the company. Great care must be taken not to single out one person within the team for a special award, if any marked achievement is demonstrably due to the activities of the whole team. Cross checking with the managers of the other members of a multifunctional or interdisciplinary team, as well as with the team leader, is essential before giving any reward to an individual.

It has in fact been suggested that the optimal incentive scheme for groups with a long-life span is one that promotes cooperation between workers through team working and peer-monitoring. Therefore group incentive schemes provide the best method for increasing the motivation of teams.

2.1.5
Disciplinary Action and Trade Union Representation

Even in the best-managed and highly motivated groups of scientists, there are occasions when disciplinary action will need to be taken with some individual.

A case of failure to meet performance criteria over a reasonable period of time will initially be dealt with by the performance related pay system. However, when poor performance persists over an extended period, even after trying the person in other roles and giving an official warning, then termination of employment will be the result. Depending on the company's procedures this could be a lengthy process and the HR manager needs to be involved at every stage to ensure that an "unfair dismissal" charge is avoided.

The same procedure will be followed if disciplinary action is required when a person fails to meet the general terms and conditions of employment, outside or beyond personal performance criteria, agreed with the company.

In many cases disciplinary actions will involve discussion with the relevant trade unions. This will depend on many factors including whether unions enjoy full negotiating or representational rights with the company. In most countries of the EU such rights are enshrined in the Social Chapter but not all countries are signatories to this part of the treaty. Negotiations with trade unions will be lead by the HR function. A Manager needs to be aware of the procedures to be followed within the company and not take unilateral action.

The last few years has seen a significant growth in the use of fixed term and personal contracts, where the reward is not governed by collective bargaining [A-4]. There are many reasons for a company wishing to use fixed term contracts and a few in R&D are given below.

- To retain staff for short period in order to complete a project.
- To create a period of review for new staff before offering permanent employment.
- To utilise experienced staff who may only want to work for a fixed period.
- To employ on a temporary basis somebody with a particular skill or competence.

Fixed term contacts are ones that run for a fixed period of time regardless of the level of the employee or whether by oral or written agreement. Having a written agreement is obviously advisable for most employees and, as a very minimum; the contract should contain the following clauses.

- A termination date, e.g. "until 16 April 2008 or for two years from 16 April 2006".
- A statement on renewal, e.g. " a decision on renewal of this contract will be made on 16 April 2008".
- A termination clause for breach of contract, e.g. serious misconduct, breach of confidentiality, inability to complete the contract due to extended illness.

2.1.6
Career Development

Most people want to see a career path stretching in front them.

Career development, like monetary rewards, is something that is of concern to everybody in employment, particularly for staff in their early years with a company. Most people in work, with the exception of those nearing retirement age, like to feel that there is a career path stretching in front them. They want to know where this is leading and how they are going to progress along the route. It is the job of any Manager to recognise this desire and to meet the need, whilst ensuring that people's ambitions are realistic. Holding out the possibility of progress to someone when this at best seems doubtful, will only store up problems for the future and managerial reputation will suffer when nothing in fact happens.

The Industries offer many career opportunities for progression and change and a variety of entry points for the professionally qualified scientist. Some examples of these are listed below.

- New Product Research
- Analytical
- Process Development
- Technical Service and Marketing
- Plant Management and Production
- Health, Safety and Environment
- Regulatory Affairs
- Intellectual Property
- Information Services

In larger companies many of these disciplines still exist as separate departments or functions though increasingly aligned with individual business units. In smaller companies this is not possible and several of these jobs or responsibilities will often be rolled together under one manager, or outsourced to another company or consultant. As an example, in a smaller company it is unlikely that an individual will be employed to deal solely with the intellectual property issues of patents, copyright etc.

The professionally trained chemist working in a research function will be faced with a career dilemma. Carrying out research is an enjoyable and intellectually rewarding experience. Going into management to reap a higher financial reward and wider career opportunities will mean that this enjoyable activity has to be sacrificed. This move could also mean the loss of a dedicated and able researcher to the company and in order to try and resolve this conflict many operate what is called " a dual ladder system". The dual ladder is a method by which top scientists can stay within the R&D function, but in a non-managerial role, and still receive similar financial rewards. By this means companies are able to keep their top scientists carrying on with their research, without any feeling of frustration over lack of reward or career progression.

There are several questions that a Manager might ask of individuals who are at a turning point in their careers, to pose to themselves relating to the science versus the management pathway dilemma, [A-21]

- *Do I like to focus on team or individual results?* An academic leaning tends to focus on individual results
- *Do I communicate effectively?* Communicating scientific results   to one's peers is one thing but convincing others of a plan of action, e.g. the board of directors, is quite another thing.
- *Can I give constructive criticism?* Criticising constructively others scientific results is okay but giving the same to non-scientific staff is a different skill.
- *Will I be comfortable giving orders and constructive criticisms to my former peers?* This

puts a special strain on friendships, as a supervisor your colleagues will never view you in the same way again.

- *Will this bring stress into my life, and is that okay with me?* There is a great deal of difference from being anxious about the results of experiments or personal goals to the stress from top management when a project timeframe slips.

Therefore, the performance management can help individuals to seek answers to questions on their career aspirations. Staff can be helped to think about alternatives for which they may be suited and thus their horizons broadened. The data from these discussions should always be logged for future reference.

When people are at the beginning of their careers they need to gain experience in several areas of the R&D activity of the company. This is especially true of those people in the technician grades where the primary knowledge is being gained by in house work. For these latter people regular rotation through jobs is a good practice. For professionals, any changes in their jobs will be governed by progress on their specific R&D topics as well as their need for broader experience. Similarly, the training plans for the individual should also be matching their planned or potential career pathway. A useful technique for assisting in this process involves the drawing up of possible *career pathway maps* for each level of staff with associated training inputs.

Figure A4 shows a typical career pathway for a person entering a process technology group directly from high school at say 18 years of age. This person continues in academic education through part time study for a BSc at the local university or college or by distance learning. Job experience in parallel to this education is achieved within the company in relevant areas. Training in practical organic chemistry is achieved by working on the process development of products to be manufactured on the plant, with the supervising chemist acting as a mentor; knowledge of analytical methods is gained either incrementally, by using the various techniques and instruments, or by direct transfer into an analytical support group for a period of time; experience of plant operation is gained by helping to run processes on the pilot plant or by being employed as part of a shift team during the experimental manufacture of a product by a new process. On completion of the degree course by part time study, this person should be ready for promotion to the position of development chemist; acting in a more independent manner than has previously been the case. Experience in this role is then gained over a period of time by working on a variety products, processes and plants, including periods acting as a shift team leader. This is coupled with personal development training that is aimed towards a future managerial role. After this time has passed, which may well take a few years, the person is ready for a job in the management of a manufacturing unit or plant.

The second example of a career pathway is for a postgraduate chemist entering as a new recruit into a research function (see Figure A5). The initial job is as part of a team working on new product research, the job holder responding directly to the Team Leader. After a period of time, say one to two years in this role, the Manager has planned a move into another group so that the chemist will gain wider experience. Other moves are possible, for instance into a role involving

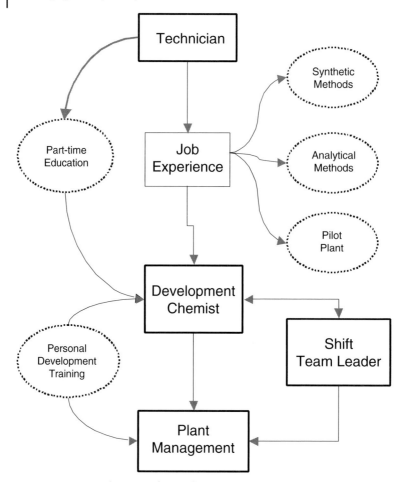

Figure A4. Career Pathway Map for a Technician

process development and hence contact with the plant and manufacturing or into a technical service job to gain experience in dealing with customers and activities at the business interface. Alongside these moves, training in both technical and personal development aspects will be provided to assist in career progression. The next career move will be into the Team Leader role. A sideways move into a marketing job to create greater commercial awareness may at this stage be beneficial. Alternatively, the management of a team with a more interdisciplinary content could gain a broadening of experience. The process of personal development goes on with attendance on higher-level courses that are specifically geared towards management. At the end of these moves and training the person is ready, when the opportunity arises, for promotion into the job of an R&D Manager.

These are just two, relatively simple examples of possible career pathways but they both involve significant promotional steps. Not everybody employed within the R&D group is going to see such large changes during their careers. For the majority of employees a plateau is reached somewhere around the chronological mid point of a career. When this happens motivation can be a problem and it is something that needs to be recognised and action taken accordingly. Fortunately, R&D people remain very motivated generally by the technical aspects of the job. Ensuring that the technical challenges remain high for those people on a career plateau is a key managerial task. Performance related pay is also designed to encourage good performance from these people, offering the potential for high rewards within the same grade or role.

The method for matching these career aspirations, or suitability, to actual opportunities involves having a managerial process for a planned succession through a portfolio of jobs. This is known as *succession planning* and is covered in the next section below.

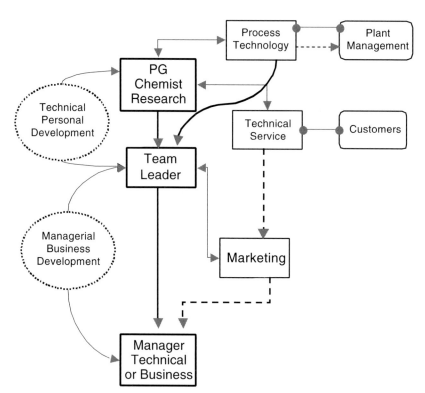

Figure A5. Career Pathway Map for a Postgraduate Chemist

2.1.7
Succession Planning

Succession planning may be almost impossible but at least make contingencies.

Succession planning is the process by which all the key jobs, in the group or department, have a person, or persons, designated as planned replacements for the current occupants who, for a variety reasons, may move out of these positions.

The controllable occasions when a successor for a job will be needed are as follows:

- Retirement
- Career move within the Department
- Career move within the Company

The uncontrollable occasions are:

- Resignation
- Medical Reasons
- Death

It is obviously very difficult to have a plan for a successor in the latter cases, especially the last two, which are often traumatic occasions. The best one can do in dealing with any uncontrollable loss is to have contingency plans for the essential jobs in the R&D group. Fortunately, R&D is one of those areas where, because of the time frame on which it operates, the loss of a person in an important position for a short period would not be disastrous for the business. The sudden removal of a key intellectual resource, a person with a specific expertise, is a serious loss to the effectiveness of the group but it does not have the immediacy of the loss of a key person in for example the sales force. The use of a competent person on a fixed-term contract can provide breathing space if required.

Succession planning is a systematic process and for the controllable moves involves the following steps.

- Make a list of the key jobs and current job holders.
- Define the timeframes for likely career moves from the career plans for these job holders.
- Identify the people who could occupy these posts on this time scale.
- Build the necessary training into the career development plans for the potential successors.

Key personnel will then be ready to move into their new jobs at the desired time with a minimum disruption to the organisation. An outline of a typical R&D succession plan is given in Table A5.

If one is not careful this process can be operated too much in a top down manner, the "manager" knows best syndrome. The career aspirations of the current staff should always be taken into account when looking for successors, a mismatch of desires can work but is more likely to lead to problems in the future.

Clearly, having a workable succession plan will depend on the size of the organisation. In a small group or medium sized company, it will be necessary to look out-side the group and discuss with other managers how a department or company can construct a workable succession plan. In many small companies succession planning may be almost impossible, and it is often necessary to resort to the employment of temporary staff from an employment agency to bridge the gap caused by even the controllable factors listed above.

This aspect of a Manager's job has become much more difficult in recent years. There has been a reduction in the numbers of people employed in almost every es-tablished chemical company and even pharmaceutical R&D has been under pres-sure. This has brought in to play a whole new vocabulary to describe this reduction in staffing levels; downsizing, de-manning and de-layering being just three examples which are now established in everyday parlance. Managers in most small companies may recognise succession planning as an ideal to aim for but one that is irrelevant to the way they are forced to work. However, it should be borne in mind that in small companies or groups the loss of key staff could be an even greater disaster than in the larger ones, as there is no spare capacity. In these companies it will be more a case of constructing a series of "what if" scenarios and making contingency plans.

Table A5   Succession Planning

Job & Holder	Career Move	Timing	Successor
Group Leader Dr J. Williams	R&D Manager	18 months	Dr A. Smith Currently in Process Technology for career development
Polymer Processing Technician Mr B. Brown	Retires	2 years	Ms. J. Jones Move to work with Mr Brown for mentoring during the next year.
Technical Service (Polymer Additives) Ms. C. Peters	To a job in Marketing	1 year	Mr S. Adams Gaining experience dealing with customer problems. Needs training in presentation skills

3
The R&D Team Manager

Team working is key to running a successful R&D Group.

Team working has been seen in recent years as the key to running a successful business. A whole consultancy industry has been built on the supply of training and help to managers in Team Building. Using this as a search term on the Internet sites of companies such as Amazon shows the large number of textbooks there are on this topic. There are many types of teams described in the literature but those of interest in the R&D arena include the following.

- Self-managed (empowered). Whilst a consequence of the flatter management organisations that exist in most companies. These types of team are now the cornerstones of the management of many R&D projects. Self-managed teams are designed to improve flexibility, drive customer service improvement, speed up product delivery and maintain motivation.
- Cross-functional. Commonly used as project teams to develop and launch new products where integration of R&D, Production and Marketing functions are required. They are designed to overcome the vertical barriers that develop between functions.
- Multi-disciplinary. These are key to the developments of newer technologies within R&D and involve the pooling of expertise of different professionals.
- Trans-national. Working across different cultural boundaries is a common requirement with increasing globalisation of the industry. Expertise in one country is linked to another in a different country across time zones, aided by fast telecommunications and networks. Problems to be overcome include cultural and language differences.

In their quest for the holy grail of successful team management in R&D, new and even old managers can find this variety quite confusing. Their only consolation is that nobody has the complete answer for what to do in all cases. The management of a sporting or artistic team, for example in the production of a play in the theatre, is a good working analogy for research team management. They both involve constructing teams around a core of very creative people with a common goal. However, as any manager of a soccer or basketball team will tell you, managing a team is a very complicated and unpredictable activity with a high failure rate, especially if you are aiming to be the best. The simplest approach to avoid failure is to recognise the

Research and Development in the Chemical and Pharmaceutical Industry, Third Edition. Peter Bamfield
Copyright © 2006 WILEY-VCH Verlag GmbH & Co. KGaA, Weinheim
ISBN: 3-527-31775-9

pitfalls and to try and not fall into them. Whatever the type of team there are some basic managerial actions that need to be carried out.

- Appoint a good leader.
- Describe the goals clearly; make clear the vision and align with strategy.
- Select the correct number of people with the right skills and competencies.
- Appoint team members who are good at working together, showing mutual trust and respect, and who demonstrate flexibility and adaptability.
- Make the necessary training input.
- Ensure that an effective method of communication is in place.

The idea that scientists are individuals struggling in isolation towards some great breakthrough is very largely a myth. The principle of team working has been enshrined within the organisation of industrial R&D for generations. However, these teams are increasingly multi-disciplinary and cross-functional, as the boundaries between the various scientific disciplines have become very blurred. This is particularly true of the life science based industries, such as pharmaceuticals and agrochemicals, where teams often consist of chemists, biologists of all varieties, geneticists and ecologists etc. Similarly in the electronics field chemical developments involve not only chemists but also material scientists, electronics engineers, physicists and system engineers.

This part of the book is restricted to a consideration of the skills that an effective R&D team manager will need to deploy. These are specifically leadership skills, the management of creative groups and last but not least the ability to manage one's own career. The various additional aspects of team working, especially the organisation of project teams will be examined in greater detail in other Sections.

3.1
Managerial Skills and Leadership

The identification of the component skills that make up a successful leader has been the subject of extensive research in many schools of social science and psychology etc. What is clear from all this research is that natural leadership skills, or the "born leader" qualities, do exist but that they can only be developed not implanted in a person. It is possible, however, to identify other qualities that can indeed be taught to individuals.

In terms of Managerial Leadership Skills these qualities have been identified as shown below [A-13, A-14]. Those qualities which are marked with an asterisk *, are the skills which can be taught to an individual and improved by practice and coaching.

Perceptual or Conceptual Skills
- Information Search*. Gathers different information from a wide variety of sources to help in decision making.
- Concept Formation. Creates strategies, builds models and forms hypotheses and pictures on the basis of information.

- Conceptual Flexibility. Identifies options and feasible alternatives, evaluates pros and cons and is willing to abandon or change direction.

Interpersonal Skills

- Interpersonal Search*. Seeks to understand ideas and feelings of others, can see things from another person's viewpoint.
- Managing Interaction. Is an effective team builder, makes team members feel valued and empowered.
- Development Orientation. Encourages the development of individuals, provides coaching, training and other resources to improve performance.

Presentational Skills

- Impact*. Uses a variety of methods, e.g. persuasion, alliances, networking, appeals to the interest of others to gain support for ideas and strategies.
- Self Confidence. Confident and unhesitating decision maker and action taker, confident in the future success of actions.
- Presentation *. Presents ideas clearly, both orally and written, enables others to understand the communication.

Motivational Skills

- Proactive Orientation. Structures the team and takes responsibility for all aspects of the team, including success or failure.

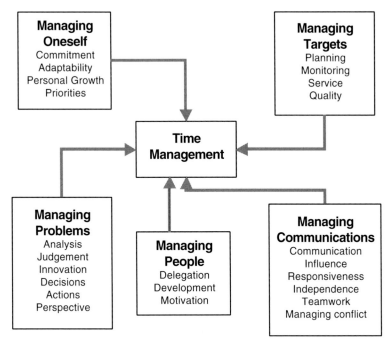

Figure A6.   General Management Competencies

- Achievement Orientated. Has high internal work standards and sets ambitious but attainable goals, wants to be more efficient and to measure progress towards a target.

Using this system it is possible to rate personal skills in each of these categories. It is generally reckoned that a successful manager is very effective in four of these eleven skills, is effective in another four and ineffective in three.

General management competencies can be split into six areas.

1. *Managing Oneself.* Directing your own efforts in an effective manner.
2. *Managing People.* Getting the best from other people.
3. *Managing Communications.* Dealing with others in a range of interpersonal situations.
4. *Managing Targets.* Ensuring that objectives are achieved.
5. *Managing Problems.* Understanding information and concepts in order to take effective decisions.
6. *Managing Time.* Ensuring that time does not become an issue for oneself or the group.

The individual skills making up these typical competencies are illustrated in Figure A6.

3.2
Managing Creative Groups

Appoint charismatic people rather than technical superstars to lead creative teams.

The essential difference between R&D and other functions within the company is that its sole reason for existence is to be creative (see also Section C). Hence any successful R&D Manager must be able to manage creative groups.

The management of creative groups is a little studied topic but it obviously requires a Manager to use the basic skills, as defined above, in a different manner than with other, less creative groups of people. Once again for the scientist it is useful to look for analogies in other creative areas. An interesting contribution comes from a Creative Director of a large advertising agency [A-15].

> "Creativity's norm is not order but chaos because the results and the time they take are totally unpredictable. Therefore, creative people are insecure; they never know where or when the next idea is coming from. So, they need support, affection and permission to fail. The creative mind seems very much associated with a sense of play, fun and childishness. It does not like discipline, particularly that imposed by the logical ("dull, slow") mind. Rewards, salary, bonuses etc. are not so much needed for themselves but signs from you, the boss/daddy, that you value/approve/love them for themselves and at least as much as their colleagues/siblings. They need things to be fair, they are acutely aware of bias. The allocation of opportunity and the judgement of their work must be absolutely

above board. The person judging their work must be someone they look up to either because he or she has done great work themselves or because they have a reputation for inspiring great work. Creative people need to know they are improving, and that you are helping them to reach higher goals, for which you must never take credit. The organisation for which they work should for maximum benefit, have high ideals to which they can subscribe. If the company is just in it for the money and cops out on their work, for example, they feel it deeply; they become demotivated and cynical".

The messages given in this quotation will strike a chord with anybody who has been involved in managing an R&D group in the Chemical and Pharmaceutical Industries and elsewhere. The main thing about creativity, as pointed out above, is that it is quite unpredictable and not guaranteed. For many people who sponsor industrial research failure is regarded as a luxury, and hence there is often a great deal of insecurity in people who are expected to be creative, the "fear of failure" factor. The R&D Manager must make creative people feel that they will not be punished for failing and must install a climate in which such fears about failure do not exist, a point emphasised above and elsewhere in this book. Remember people with ideas that fail at least are people who have ideas.

The reward systems, which have been dealt with earlier, can be misused with creative people. It is very easy for them to feel that only concrete measures of performance will bring rewards and that the more intangible elements, needed for creative work, will not be given their just value. They will often only respond to value judgements on their creative performance made by somebody whose own creativity, or ability to inspire creativity in others, has been proven. A Manager who has no track record in research has to work hard to get the wholehearted support and respect from high performing researchers.

Therefore an effective Manager of a creative group typically demonstrates the following characteristics:

- Is able to create the right atmosphere which allows for uncertainty and failure. This atmosphere is highly supportive but rigorous over technical competence.
- Is fair and equitable in the use of power, ensuring fair treatment of all staff by not just rewarding those whose work is immediately successful.
- Studies to be technically competent, and gain the respect of the researchers. Is seen as "one of us" rather than "one of them".
- Is visionary and enthusiastic whilst keeping both feet on the ground in order to ensure progress towards the goal.
- Is sympathetic to the scientific goals of some creative people when terminating a project, e.g. allows just enough work to be done to complete work for publication before work ceases.
- Is tolerant of personal idiosyncrasies often associated with creative people. However, poor quality work is not allowed since there is a high concern for standards.
- Is able to live with ambiguity, for example is able to live with chaos whilst requiring stability and with failure whilst needing success.

Other studies have given support to these managerial characteristics and further emphasised that creativity amongst scientific personnel is more likely in teams where there is a favourable relationship with the Manager [A-16]. This is important in producing the very valuable, non-material benefits that are listed below.

- Greater autonomy allowed for researchers
  - Freedom to try out new ideas
  - Allowed to undertake non routine tasks
  - Able to work on personal projects
- Greater managerial support (emotional and administrative)
  - Support for risky projects
  - More motivating and encouraging
  - Quicker to act on financial/equipment requests
  - Less likely to penalise failure
- Excellent relationship between Manager and staff
  - Staff more committed to the organisation
  - Staff show higher work satisfaction
  - Staff have better attitudes to innovation
  - Greater creative output from the team

The same studies indicate that it makes sense to appoint charismatic people rather than technical superstars to lead creative teams. Technical expertise is important in R&D Managers but not as essential as being able to inspire, motivate and energise research staff. Charismatic leaders demonstrate the following characteristics.

- Inspired visionaries
- Have high interpersonal attractiveness
- Can create a sense of urgency amongst scientists
- Generate high levels of commitment from staff to the company
- Foster creativity

Compatibility between the Manager and researchers is important. For instance having a similar level of education. Shared values about their work, its dignity and the role of rewards also help to develop a common language.

Even armed with all this information the selection of the right person to act as manager of a group of creative people within R&D is no simple matter. Within R&D, depending on the nature of the work, differing degrees of creativity are expected from the component groups. The work within a large department can range from highly speculative research to projects with a fixed deliverable or to specific customer orientated short-term work (see Section B, 1.2). The managerial skills required across this range of activities will vary widely. Management of the latter require both general and project management skills whilst the former requires the special skills needed for managing creative people.

The difficulty in selecting managers for highly creative groups has lead some people to question whether a manager is needed at all for such groups. This would be a classic case of self-managed teams, which on the face of it seems very attractive. However, experience shows that whilst this is indeed workable it is not the ideal position. As emphasised above, there is a vulnerability associated with highly creative people that means they need the support and encouragement from an interested individual, even sometimes "a shoulder to cry on". Therefore it seems there must always be a hand on the tiller, but this need only be a light one.

3.3
Managing Your Own Career

Don't forget to manage your own career.

As organisations have become flatter with de-layering, downsizing and re-engineering reducing the management steps in many organisations, leading to a possible plateau for individuals in mid career, career management by the individual has become even more important than previously was the case. Additionally, the time when chemists and other scientists could join a company and expect a career for life is now past. It is highly likely that most chemists will change their job or company at some stage of their professional life.

A person cannot always rely on the good practice described above being followed by their own line managers, no matter how enlightened they maybe. For a Manager the management of their own career needs to have the same care and attention paid to it as is applied to that of any other member of the group.

Career management is something that needs to be done continuously and not just at critical moments in a person's career [A-17]. To meet this challenge some career rules for professionals have been formulated [A-18].

- Keep an open mind, be prepared to switch directions or career paths.
- Failure is inevitable. Present your failures as positive ventures, as examples of risk taking.
- Learn to recognise opportunities even when they come from unexpected directions.
- Take charge of your career, it reduces stress even if you do not make it to the top.
- Time is on your side as organisations become flatter the route to the top is shorter.

What practical actions can sensibly be taken by the individual? The key is to have a plan which identifies both short and long term goals [A-17, A-22].

- Review your current performance and identify areas for improvement. The outcome of 360° and upward assessments are ideal for looking at other people's perception of you and your performance.
- Having identified areas for improvement set yourself targets for the coming year, reviewing progress at regular intervals.
- Decide what training you require to improve your skills and competencies. Ensure that these requirements are discussed and agreed with your line manager and that money is available in the training budget.
- Have a longer-term career development plan. If your company runs assessment centres ensure that you go through the process.

- Develop networks and look out for job moves that broaden your experience and fit in with your career development plans.
- Keep your eye open for both internal and external job opportunities.

What does this all mean for people who want to have a career that leads to the top in R&D? It certainly means that they will need to have had experience in a wide range of R&D activities, in more than one area or business, probably in more than one company and preferably in more than one country. Experience in the marketing function will be increasingly beneficial in the customer driven world. Young managers will need to get involved in those cross functional working groups engaged in the so called "change programmes" based on any new HR thinking, some recent examples being empowerment and self managed teams, to raise their profile and be at the heart of topics which attract senior managers attention.

Whilst having experience in jobs outside of R&D is going to look good on a persons CV, when aiming for a career in R&D it must not last too long. These types of jobs should ideally be in the nature of a secondment in which the job holder is gaining experience. If a Manager is away from R&D for an extended period of time they may be regarded as no longer suitable for technical management, especially in a rapidly changing area. Hence there is a danger of being overlooked in the succession plans operating in a large company or, even more likely, rejection by a selection panel when applying for an advertised R&D post in another company.

References

[A-1] PRAHALAD, C.K., HAMEL, G., *Core Competencies and the Concept of the Corporation*, Harvard Business Review, May–June, **1990**

[A-2] PRAHALAD, C.K., HAMEL, G., *Competing for the Future*, Harvard Business School, Boston, **1994**

[A-3] An example is given in www.cognology.com.au

[A-4] FOWLER, A., *Get More and More Value from your People*, Institute of Personnel Development, London, **1998**

[A-5] JONES, S., *The Headhunting Business*, Macmillan, London, **1989**

[A-6] ARMSTRONG, M., BARON, A., *The Job Evaluation Handbook*, Institute of Personnel Development, London, **1995**

[A-7] TOPLIS, J., DULEWICZ, V., FLETCHER, C., *Psychological Testing: A Manager's Guide*. 3rd edn., Institute of Personnel Development, London, **1997**

[A-8] http://www.psychometrics.co.uk

[A-9] WOODRUFFE, C., *Assessment Centres*, 2nd edn., Institute of Personnel Development, London, **1994**

[A-10] NAISBY, A., *Appraisal and Performance Management*, Spiro Press, London, **2002**

[A-11] KENNEY, T., *The Observer*, London, 10th September, **1994**

[A-12] *360 – Degree Assessment; An Overview*, US Office of Personnel Management, **1997**; http://www.opm.gov/perform

[A-13] SCHRODER-COCKERILL, 1992 as discussed by J. HUNT, Royal Society of Chemistry meeting, London, **10.3.93**

[A-14] SCHRODER, H. M., *Management Competencies- Key to Excellence*, Kendall Hunt, Iowa, **1989**

[A-15] BRIGNALL, T., paper at *Managing Creative Groups Colloquium*, The Royal Society, London, **1992**

[A-16] BASSU, R., GREEN, S. G., R&D Creativity: Manager's Relationship with the Researcher is Key, *R&D Innovator*, 3 (11), **1994**

[A-17] BOCHARDT, J. K., *Career Management for Scientists and Engineers*, Oxford UP, Oxford, **2000**

[A-18] CLUTTERBUCK, D., DEARLOVE, D., *Routes to the Top*, Kinsley Lord, **1995**

[A-19] Example is www.rsc.org/chemistryworld

[A-20] CHE, Y., YOO, S., *Optimal incentives for Teams*, American Economic Review, **2001**, 91, 525-541

[A-21] JENSEN, D., www.sciencecareers.org, December 17, **2004**

[A-22] COHEN, C., M., COHEN, S., L., *Lab Dynamics: Management Skills for Scientists*, Cold Spring Harbor, New York, **2005**

Section B
Organising for an Innovative Environment

The environment in which research operates has changed over the last few decades with an increasing speed. The mid–20th century saw the high point of the strong, functional, stand alone R&D with influential membership on the board of directors of a company. It was the era when R&D in the chemical industry produced major innovations, exemplified by polymers and man-made fibres, ready to be received by an expectant company for launch onto a receptive consumer. By the late 1960s, when every major company had such an R&D function, the consumer had become more selective and hence the marketing of products became the key differentiator. At this point research was put under greater control by the board of directors. It was set specific business targets by marketing and the process began to be managed using project management techniques. In the 1980s the perceived needs of the customer became all important, the businesses became "customer-focused". The role of R&D needed to be clearly defined within an overall strategic plan for each business and innovation managed by a team from the breadth of the organisation. To differentiate it from the two described above, this methodology was called third-generation R&D [B-1]. Whilst this was fine for driving research along perceived customer needs it was deficient in dealing with step-change innovations. The importance of knowledge to any company was formalised in the 1990s by the discipline of knowledge management. R&D, being a key knowledge resource in the company, needed to be much more involved in sharing this knowledge [B-2]. Developments have become much more collaborative with both other companies and customers, especially when testing out discontinuous innovations coming from R&D. This knowledge-based methodology has been called fourth-generation R&D [B-3]. Increasingly R&D in major companies is operated on a global basis requiring Managers to be responsive to a new set of skills.

Managers are rarely free agents when they are considering the best way to organise their particular area of the R&D working environment. Many factors need to be considered, a number of which are beyond their direct control, the simplest and most obvious being the size of the R&D function or group. Whilst the functional Manager's views will be sought on the size of the research team needed to work on a desired portfolio of projects, in the end the executives of the company will set an expenditure limit. The amount of money allocated for research will depend on the nature of the business, the sector in which it operates, its size and current or projected profitability. Even allowing for this limiting parameter of scale, there are still a variety of ways in which R&D can be organised internally, each of which has its own positive and negative features. New R&D Managers, in particular, must understand the various forms the organisation of R&D can take, so they can adapt their management styles to suit the environment and culture in which they find themselves. In this Section, four of the main organisational structures are considered. These range, at one extreme, from the one based purely on the stand alone R&D function, to the other extreme where Research is carried out independently within each Strategic Business Unit with little or no functional organisation outside these Units. The middle ground is where the R&D function contributes to a fully integrated, cross-functional matrix or as a contractor to a Strategic Business Unit (SBU). The internal organisation of the R&D discipline will vary from one company to another and, in the large corporations, in the different segments of the business.

Organising the provision of the support services required by research staff is another key task for an R&D Manager. Crucially the decision needs to be made as to whether this support will be provided by an internal resource or contracted out to an outside agency, outsourcing, as it is known. Outsourcing of non-core activities is very common but, as with all activities, there are pros and cons and making the wrong decision in a particular area could lead to a competitive disadvantage. The most significant support areas to be considered in R&D for outsourcing are analytical services, intellectual property, covering patents, licensing, etc., information and library services, IT such as office technology, environmental services, toxicology and regulatory affairs. Drug development, even in major pharmaceutical companies now relies heavily on outsourcing to get a product through the regulatory hurdles.

Analytical laboratories have absorbed developments in the automation of laboratory equipment that has markedly influenced the way that they are managed for over thirty years. In the last ten years, automation has spread to the synthetic laboratories of chemical companies and is also making a big impact on the way both product research and process development are carried out. These developments have consequences for R&D Managers, which are discussed in this Section.

An R&D line Manager's personal working environment will normally include some form of financial responsibility, usually the control of internal budgets and manpower costs of research plans. Since in any R&D organisation the budget will be dominated by payroll costs a very important element in exercising control is the efficient utilisation of staff. Some guidelines are provided on this topic. Financial aspects are clearly important in an industrial environment but, from legal, moral and ethical standpoints, those relating to health and safety should be as equally high on the agenda. Whilst some regulations affecting R&D laboratories and chemical plants are voluntary others are statutory and an understanding of these is essential. The operation of a quality R&D Group should be the goal of every R&D Manager. The formal aspects involved in operating under quality procedures, and how this fits in as part of Total Quality Management methodology, is outlined in the last part of this Section. Finally, guidelines are provided for managers involved the management of changewithin their organisations.

Overview

Adoption of management styles to suit the environment and culture of the company requires an understanding of the different organisational structures. Functional, strategic business unit, R&D contractor or matrix, they all have a place. Corporate laboratories are reappearing, but with a new role. Being part of a business unit increases your influence on its strategy. New Business groups are there to handle high risk business opportunities. A matrix is most effective in the time management of cross-functional projects. Team Leaders play an important role in upholding the reputation of the function in the company. The globalisation of R&D means that an additional set of managerial competencies is required. Outsourcing is now an essential component of business strategy having implications for the management of R&D. A manager must ensure that the quality of any outsourced services is of the right standard and especially that scientists have access to the widest range of high quality information that is affordable. Developing a strategy for automation in the laboratory involves making sure that it meets the needs and objectives of both the laboratory and the company. The working environment must be financially sound, healthy, safe and organised in a quality manner. All members of staff have a general responsibility in health and safety legislation but R&D management have specific responsibilities. Regulations covering the operation of a laboratory and in manufacturing are mandatory in the pharmaceutical and food industries. TQM in no way conflicts with the long-standing goals of an effective R&D Group. The process of changing from one state to the other state needs to be structured and well organized.

1

The Structural Components of an R&D Organisation

Adopt management styles to suit the environment and culture.

Since the organisation plays a crucial role in stimulating creativity and innovation it is important to understand the nature of organisations, especially their culture, before being able to understand why some are more receptive to creativity, innovation and hence change.

Some general comments on organisations are as follows.

- Organisations often change their cultures as they grow – but may not be aware of all the implications.
- The actual culture at any one time will depend on factors like origins, subsequent history, purpose, environment, ownership, and values.

Several types of organisational cultures have been identified.

- *Power orientated web.* Power centralised in an individual, often the founder of the company. They are informal, can respond to change quickly, suit the ambitious person.

Table B1 Role and influence of Cultures on Creativity and Innovation

Structure	Role/influence
Power oriented web	Innovation governed by boss's attitude
Greek Temple	Good for competence building and nurturing. Blue sky work can be protected
Task oriented	Excellent for targeted research and getting innovations to market. Creativity restricted to short term
Person oriented cluster	Useful for evaluation of innovations
Change oriented	Identification or market led research targets. Innovations based on applications research.
Spread sheet	Outsourcing of R&D services

Research and Development in the Chemical and Pharmaceutical Industry, Third Edition. Peter Bamfield
Copyright © 2006 WILEY-VCH Verlag GmbH & Co. KGaA, Weinheim
ISBN: 3-527-31775-9

- *Greek Temple.* Each pillar in the temple is a separate function. Works by logic and procedures and carefully delegated authority. Slow to react to change.
- *Task orientated.* Emphasis on getting the job done. Typically project teams used to get a particular end result. Empowerment and flexibility motivating to the individual.
- *Person orientated cluster.* Emphasis on personal skill, e.g. in partnerships of professional people, Used to complete a specific task and then disbanded.
- *Change orientated.* Seeks greater involvement with customers and suppliers. Responsive to change needed to deal with outside factors.
- *Spread Sheet.* Groups of people with varying degrees of linkage with the organisation are clustered around and loosely coupled with the core group. Involves subcontractors, outsourcing companies and temporary staff.

The role and influence of these various cultures on creativity and innovation are shown in Table B1.

Large companies by their very nature are not uniform and exhibit more than one culture, from department to department or function to function. Organisationally each department may appear to start off looking similar but a sub culture develops that defines the management style. A new manager coming into a department needs to recognise the existence of a sub culture and understand that it might make it quite difficult to bring about change.

Consideration of the component parts of any R&D grouping therefore needs reference back to the way in which the parent company is organised. The structure that is adopted is dependent on the size of the company and the nature of its business. Most large companies, at least on paper, have adopted R&D organisations that look very much alike. As indicated in the introduction to this Section the organisation of R&D has not remained static over the years. However, change has been more evolutionary than revolutionary and larger organisations have tended to change in very similar ways to each other. In the smaller to medium companies, known as SMEs or Small to Medium Enterprises, especially in evolving areas of such as biotechnology, there is a much greater diversity. In small chemical companies this diversity is often driven by economic necessity rather than by a strategic choice.

1.1
Organisational Environments for R&D

Functional, strategic business unit, R&D contractor or matrix, there is a choice.

Only four possible organisations for R&D within companies are considered in this book. Companies have adopted many alternative structures, but these are usually substructures or variants of the four generic types shown in Figure B1. Organisational arrangements such as these are only directly applicable to those concerns that have medium to large sized research organisations and not to small companies. The organisational arrangements used in companies with few people in an R&D role are not so easily described and they must be considered case by case.

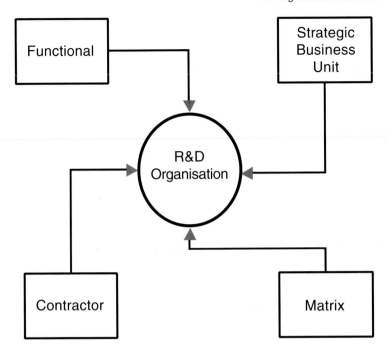

Figure B1. R&D Generic Organisations

1.1.1
The Functional Organisation

The most long-standing organisational structure within medium to large companies is that based on the functions, R&D, Marketing, Production, Finance and Human Resource. Each of these functions contributes to the operation of the whole company but is managed independently and has its own internal life; an example of the Greek temple organisation. Traditionally, all of these functions were fully represented on the Main Board, for instance by the R&D Director (Figure B2).

However, the fashion has now changed in many companies to having a much smaller Executive running the business at a more strategic level. It is therefore much more common to have a layer of General Managers, responsible for the day to day running of their function, reporting into one or other member of the small Executive. This is shown schematically in Figure B3. In these circumstances the person representing R&D may not have any technical background. This lack of direct representation of the technical functions on the Executive can be a problem for senior R&D Managers. It certainly requires senior managers to spend a considerable amount of their time explaining and defending the case for R&D spending.

There are disadvantages with this formal organisation by functions. In the worst case it is possible for the functions to become too independent, overseen by Managers with their own, separate agendas. The structure shows a strong tendency to be-

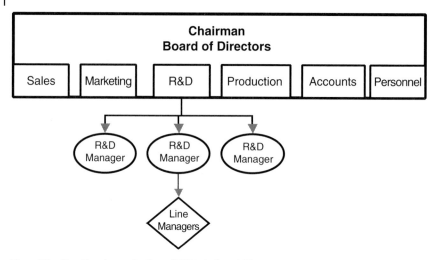

Figure B2. Functional organisation of R&D via Board Director

come hierarchical, often promoting from within the function, even at very senior manager level, with little cross fertilisation. The function, as with any closely-knit group, creates its own culture and value judgements, regarding all outsiders as the causes of its problems.

A big disadvantage for staff below the Manager level within the R&D function is that they mostly stay within their own environment, with a tendency to become isolated from the activities of their colleagues in the broader business. This can lead to missed opportunities for new business ideas, constructive appointments to job va-

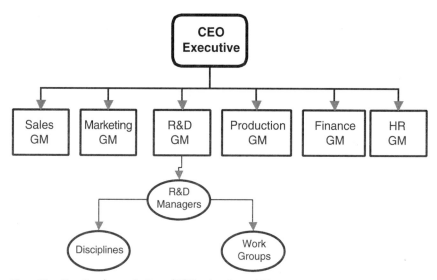

Figure B3. Functional organisation of R&D using General Managers

cancies and career development opportunities, on both sides. Potentially the R&D function can become divorced from business objectives and be slow to respond to changing business needs.

In spite of these many drawbacks, the functional structure does have distinct advantages for the maintenance of professional competencies and the development of new ones, with a view to the future R&D needs of the business. It represents a stable and friendly home in which the uncertain life of the researcher can be nourished and developed. Having all R&D under one roof creates a critical mass that allows minor, but important activities to exist, something that is impossible in the more devolved and smaller organisations. The R&D function also has a higher capability when it comes to the management of strategic research, that which is needed for the longer-term development of the business. It is also better at looking beyond the horizons of today's business for new avenues and new directions for the company.

1.1.1.1 Corporate R&D

Corporate laboratories are reappearing but with a new role.

Central research laboratories are the ultimate in the functional organisation of R&D within major companies. Often known as Corporate Laboratories, funded directly from the main board and responding directly to the R&D director or president, they are sheltered from the day-to-day vicissitudes of the individual business fortunes. For many years they have been the home for exploratory or "blue-sky" research and fundamental research regarded as important for the future development of businesses. However, they are a direct cost to the individual businesses and during the late 1980s and early 1990s, the period of cost cutting and company reengineering, many were either reduced in size, or even closed down altogether, in most companies in US and Europe. This was largely due to the increased emphasis on the Strategic Business Units (SBU) within companies (see Section B, 1.1.2). These SBUs did not want to carry the overheads that arise from a central research group. With the recognition of the importance of knowledge and its development within companies the total demise of a central research function is now realised to have been wrong and many companies are revitalising their corporate R&D efforts. However, this is not just a simple case of them returning to carry out basic research and their old function within the organisation. They are still being given independence from day to day business and extended time frames to carry but with the following new roles [B-4].

- To leverage technologies and platforms
- To house critical competencies and expertise
- To create new technology
- To fuel growth and business development

An example of this rethinking was provided by Dow, which, at the time employed 75 % of its R&D personnel within business units and 25 % in corporate R&D.

In the late 1990s the company reorganised the part of R&D effort used by all businesses into nine global technical capabilities that it saw as required for the

21st century. Corporate R&D became responsible for the new science that leads to new technology and then the ability to make it to the market place. Corporate R&D was where the world-class expertise would reside that no single business could afford.

Nine capabilities
1. Analytical sciences
2. Catalysis
3. Biotechnology
4. Materials research and synthesis
5. Computing, modelling and information services
6. Engineering sciences/market development
7. External technology/intellectual asset management
8. Materials engineering centre
9. Health and environmental research

Four platforms for growth
1. Advanced material for electronics
2. Bio-materials and bio-processes
3. Advanced surface treatments
4. Post-polymerisation catalysts

Rather interestingly Dow and also DuPont have very recently changed tack slightly and formed bodies to oversee their company's research portfolio; a "Research Council" in the case of Dow and a "Growth Council" in DuPont. [B-36, B-37]. DuPont have described their system as follows.

Corporate R&D (CR&D) is the foundation of our science efforts and has been responsible for most of our major product breakthroughs. CR&D provides both leveraged scientific services to the corporation and long-term research activities. In the past two years, CR&D has transitioned to an enhanced research model that has significant benefits to the corporation, including:
1. Long-term research focused on providing options and creating new revenue.
2. Creation of the Growth Council to select projects and oversee our portfolio to assure growth.
3. A research review process to ensure that project goals and business cases are reconciled regularly.
4. Leveraged services that provide cutting-edge technology and support to the corporation.
Growth Council Research
In 1998, DuPont began a transformational change in its approach to long-term, discovery research. DuPont changed from an organization based on scientific competency areas meant to provide discoveries for DuPont to commercialize to the Growth Council research process. The Growth Council process starts with proposals submitted for research projects and must contain a technical and business case for the research to be pursued for DuPont. Each of the proposals is evaluated by a group of senior business leaders within DuPont (Growth Council) who

determine if the proposal meets the criteria to resource the proposal. Through this process we manage our long-term research as a portfolio of projects.

Transitioning to this project structure, CR&D developed processes to evaluate, activate, staff, track, develop and, if necessary, terminate projects. Growth Council ensures a continued match between the business case and the technical accomplishments of projects.

Growth Council research is divided into three areas:

1. Biochemical Science and Engineering
2. Material Science and Engineering
3. Chemical Science and Engineering

Such research councils existed many years ago in the old style chemical majors, such as ICI, BASF and Bayer. The difference now is that the people deciding the choice and future of projects are business leaders and not research managers.

1.1.1.2 New Business Venture Groups

"New Business Groups examine high risk business opportunities"

An alternative to carrying out the more speculative business orientated research within the Corporate R&D is to set up new business venture groups.

Such groups have the remit to either sponsor businesses that require R&D collaboration between existing businesses before coming to fruition or to examine new businesses opportunities that would be too high a risk for the main businesses. Such new business groups can also be responsible for licensing in technology from small start up companies or for forming new ventures with such fledgling businesses. (See also Section D, 1.3) [A-51]

A survey in 2005 on how major chemical companies were handling new business opportunities came up with the following examples [A-52].

- BASF

BASF Future Businesses has 12 people who place research contracts with one or more of the existing R&D units. This group of people are responsible for cooperating with start-up businesses, working with industry partners, potential customers and building relationships with relevant universities. A subsidiary BASF Venture Capital can invest capital in start-ups to pump prime new developments.

- Degussa

With 90 % of R&D being carried out in existing business units, the company set up a group, called Creavis, which is charged with looking at projects whose tine frame is four to eight years. It is also able to help fund higher risk projects within the existing portfolio of businesses.

- DSM

DSM Venturing & Business Development was set up to enable the company to focus on future high growth markets. Whilst it is business oriented it is also concerned with two-way licensing technology both within the company and also venturing activities across the company (See also Section C, 3.2 – 3.4).

What does this mean for the individual Research Managers working within an existing Business Unit? Well there is the danger that they will see it as somebody else's job to have longer-term ideas. This has to be avoided and they should encourage their researchers to think beyond their current confines and then be prepared to take any bright ideas to the New Business people with sufficient data to allow the to consider the merits of the proposal. Presentation and negotiating skills will of great benefit on these occasions. Additionally, because such managers may be required to put some of their own groups effort into cross business projects, experience of working with a matrix is vital (See 1.1.4)

1.1.2
The Strategic Business Unit

Being part of a business unit increases your influence on its strategy.

In the 1980s there was a big shift from running the whole organisation by functions, to one involving the devolution of power into specific business groups, which became known as Strategic Business Units (SBU). In this organisational method, the overall business is segmented into a number of smaller units. These are those businesses, considered to be strategic in the future development of the company, grouped on the basis of product lines, markets and customers. Each SBU is given the responsibility of managing the functional components necessary for its development, including R&D. This structure, sometimes known as the business tube model, is shown schematically in Figure B4.

Potentially there are major problems for R&D within the overall company, especially if the SBUs are of modest size. These problems are in essence the converse of the strengths of the functional organisation. For instance, in the smaller SBUs it is very difficult to build up the necessary skills base for carrying out any effective R&D. The problem revolves around having a critical mass, large enough to justify the required infrastructure. For example, it is difficult to put a person with a specialised skill into team on a part time basis or for a specific short period of time if that person has not got the support of a wider function; a home to go back to on completion of any secondment. Career planning in R&D is also made more difficult unless there is a good working career development process operating across the SBUs. Certainly strategic research, work that will not be applied for at least five years, based on core competencies is almost impossible. The managers working in the SBUs are geared to meeting the challenges of today and, in R&D terms, this means getting competitive new products from R&D out to meet customer's current needs, as quickly as possible. Whilst this is highly desirable for business profitability, there is a danger that in these circumstances R&D becomes very short term and development orientated, with the search for genuinely innovative new products or processes taking a back seat. In the worst cases, the time is never right to put some speculative money into R&D. The future is mortgaged for the present. This has been called "the tyranny of the SBU" [B-5].

In spite of these potential difficulties, the big advantage of the SBU is that an R&D

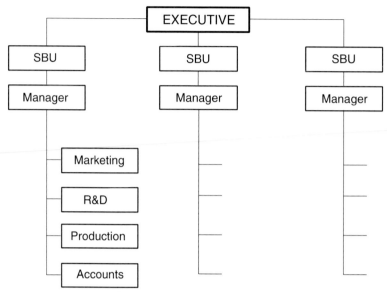

Figure B4. Organisation by Strategic Business Unit

Manager is genuinely part of the business team and is therefore directly aligned with its policy. This position can be used by the R&D Manager to influence the thinking of the business team. One result could be that any difficulties that could be faced over the development of the skills needed for the longer term are minimised, or even removed. An additional benefit of the SBU is the intimate relationship R&D personnel have with their colleagues across the business unit. They can really see that they are making a direct and measurable contribution to the welfare of the business as financial data and business performance information is shared.

1.1.3
The R&D Contractor

Another way of overcoming some of the difficulties that can arise with the SBU organisational structure is for R&D to be outside the business tube, acting as a contractor to the business unit. This can be described as "insourcing" as opposed to outsourcing In this case, R&D retains its functional independence, contracting out its services to the business unit, paid for on an "as needs" basis. The work is carried out physically within R&D laboratories on behalf of the business unit. This is illustrated in Figure B5.

Management of R&D is now off-line to the SBU units and responds directly to a mentor on the Executive, who ensures that innovation stays on the company's agenda. The R&D Manager is responsible for generating sufficient contracts, or Service Level Agreements, with the SBU managers to cover the overall annual costs of running group, which have been agreed with the executive. The day to day management of the staff, covering recruitment, professional development, training, career planning and reward remains in the hands of the R&D Manager.

This system makes it possible for the R&D Manager to overcome the critical mass problem, outlined in Section B, 1.1.2. Some specialised skills and competencies can be retained and developed, that could not be justified by a smaller group within an SBU.

> For instance, a business wishes to undertake a specific development, which re-
> quires a polymer chemist to meet the technical needs of the programme. It is
> envisaged that the programme will last for less than one year. At the end of this
> programme there will not be any further work for the polymer chemist within
> the business unit. This programme would therefore be difficult to undertake
> within a small business unit. Operating in the contract mode, R&D supplies this
> resource from the pool of technical skills available within the group. The pro-
> gramme of work is then undertaken within R&D, in collaboration with the busi-
> ness unit. Once the contracted work is completed, it is the responsibility of the
> R&D Manager to place the polymer chemist in another programme.

Carrying out a limited amount of more speculative work in R&D is still the problem. If R&D is doing a good job in getting support in its contracting role, it is possible for it to seek financial support for some more speculative or strategic research directly from its sponsor on the Executive of the Company.

There are dangers for R&D in this contractor role, especially if the business unit management do not feel any ongoing responsibility or commitment to the people or the even the resource. Business managers may only want to place one off contracts to

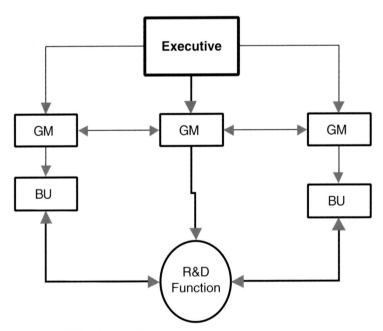

Figure B5. R&D as Internal Contractor

meet specific needs, wishing to turn R&D support, and its associated costs, on and off in response to short term budgetary controls or difficulties. To avoid this happening the R&D Manager needs to work hard in ensuring that most, if nor all, the annual agreements are based on a rolling programme of projects into the future. R&D has to be seen as having an essential place in the strategic plan of the business unit. In this system of managing and organising the R&D resource the benefits significantly outweigh the disadvantages and it has found favour with many chemical companies.

R&D staff also face problems in this contractor role, which are not insignificant. The main one is the uncertainty over the continuance of a contract and what will happen to them when a contract comes to an end. Reassurance and support from the R&D Manager is essential for the morale, not only of the individual but also of the total group. There is in addition the problem of staff seeming to have two bosses during the term of the contract, something that also occurs more obviously in a matrix organisation, and this is the next topic to be considered.

1.1.4
The Matrix

Use a matrix in the time management of cross-functional projects.

The final organisational structure to be considered is the matrix. Matrix organisations grew out of the need to have a compromise between the functional grouping responsible for competence and resource utilisation, and the product grouping responsible for coordination, control and goal accountability. There are three basic types of matrix organisation.

1. *Co-ordination Matrix*
   • Managerial authority retained by department heads (HODs).
   • Project Manager (PM) has monitoring and coordinating authority and project control.
   • PM negotiates with HODs for staffing, scheduling and change of plan.
   • PM leads the project team and links with "customer".
2. *Overlay Matrix*
   • Managerial authority shared between PM and HODs.
   • Project staff directly involved with both their professional group and the project team.
3. *Secondment Matrix*
   • Specialist staff (e.g. R&D) seconded full-time to the project team.
   • PM has full managerial authority.
   • Staff return to department at the end of their assignment.
   • HOD retains responsibility for long term training and development.

In most companies the co-ordination matrix is the one used, especially where projects involve R&D. In this matrix, as with the R&D contractor mode described in Section B, 1.1.3, scientists are bound into their various functions or disciplines for all professional and personnel matters. There is a set of Project or Programme managers, outside the functions, forming the other axis of the matrix, who are responsible for running the projects as defined by the business. In order for them to carry out this responsibility they have staff allocated to them from the various functions. It is important to realise that Project or Programme Managers do not have any line management role. At the end of the projects, the staff concerned return to work in their respective functions whilst awaiting reallocation to new projects. This structure is shown in Figure B6.

Matrix management is very effective in the time management of cross-functional projects involving the introduction of new products, processes or plants. However, it is of little value in those programmes that are heavily biased to one function, or have an extended time frame with an uncertain endpoint, typical of the more speculative R&D programmes.

The matrix brings with it the problem, already alluded to in Section B, 1.1.3, of the staff having two bosses. The personnel issues are an area for potential conflict between managers, which can impinge on the allocated staff. Recognising the different roles of each manager helps to eliminate such a problem:

- The Project Manager sets the Project objectives and is in quantitative control.
- The Functional Manager ensures the quality of the input from the individual.

Issues such as reward and promotion are the ultimate provinces of the functional manager. However, in the area of performance related pay and performance management the setting of criteria should be a collaborative effort between the two managers and the person concerned.

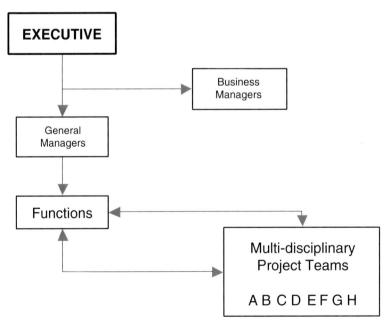

Figure B6.   The Project Matrix

When R&D supplies some of the Project Managers to the matrix there is one point that needs to be watched. It is important, if it is intended that these people should come back into R&D, that they do not stay too long in this role. If the time out of R&D is lengthy, technical obsolescence is a danger for the individual involved.

The matrix system is commonly used internally within R&D Groups, especially when research is product orientated and multi-disciplinary. This will be covered in Section B, 1.2.1. Most companies use matrix management for major projects, whether these be internal, e.g. the construction of a plant or facility, or have an external end point, e.g. the delivery of a product or service to the market. The specific use of empowered multi-disciplinary or cross functional teams working in a matrix, applied specifically to the reduction in the lead time for delivering projects, will be considered again in Section D.

1.1.5
Organisational Comparisons

From the above outline of alternative organisations clearly there is no one ideal system. The choice of organisation that is employed will depend on the role and objectives of R&D within the company. In the larger companies it is likely to be a judicious mixture of the four main types or variants that are adopted. There are three main questions to be asked and factors to be considered by R&D in adopting any structure.

1. Does it maintain the R&D skills base?
2. Is there a good connection with the business function?
3. Is there clarity of the management role?

A comparison, in these terms, of the four alternatives described in this Section is given in Table B2.

Table B2   Organisational Comparison

	Functional	*S.B.U.*	*Contractor*	*Matrix*
R&D Skills Base	Strong	Weak	Strong	Medium
Business Connection	Weak	Strong	Medium	Strong
Management Clarity	Good	Good	Good	Poor

Project organisation within a matrix allows entrepreneurial activity to flourish, providing opportunities for project managers to prove their capacity to lead teams effectively. Functional departments remain necessary as they provide the home of technological and scientific competencies and the nursery for fledgling projects.

Transferring R&D output across organisational boundaries remains a problem, as there appears to be a cultural mismatch between R&D and other functions within a company. Possible ways of avoiding this mismatch, and at the same time improving the performance of teams within the matrix have been suggested [B-6].

- Establish human bridges by rotating R&D staff to jobs in other company functions.
- Use career progression paths for young R&D staff to other functions creating networks of people able to assess new ideas and progress action plans.
- Assign a sponsor, ideally a senior executive, to a project. Their job is to follow its progress and guarantee its survival if conjunctural difficulties are causing difficulties, whilst assessing its value in a detached way.
- Establish multi-functional project teams as a temporary organisational solution.
- Joint programme planning between R&D customers and R&D staff.

As discussed in the introduction to this Section, it is possible to identify four generations of organisational methodologies for R&D within companies. This is organisational theory on the macro scale, not something that young, new managers can change, but they should have an understanding of these models so that they can evaluate where their own company stands in relation to each generation. Only a very brief overview will be given here and the reader is advised to consult the original publications for detailed explanations [B-1, B-3].

- *First generation R&D companies*
  - Executive sees R&D as a line in the annual budget, it is regarded as an overhead cost.
  - R&D decides the technologies and there is no explicit link to business strategy.
  - Unbounded search for scientific breakthroughs.
  - Seek to leap from current knowledge to new knowledge and hence to new products.
  - Technology first, business implications later.
  - Lack strategic objectives.
  - Avoid the matrix, strong functions, emphasis on cost centres and discipline.
- *Second generation R&D companies*
  - Joint consideration by business and R&D about the strategic direction for the company, but only on a project-by-project basis.
  - Supplier- customer relationship between R&D and the business on incremental R&D.
  - Fundamental research carried out in central R&D.
  - Matrix management of projects.
- *Third generation R&D companies*
  - R&D operates a strategically balanced portfolio generated co-operatively by R&D and general managers.
  - Partnership operates with business so that R&D is no longer isolated.
  - Defined and consistent business objectives set for R&D
  - Use market surveys to determine existing customer needs and targets R&D to meet these needs.
  - Emphasis on continuous innovation hence limiting on discontinuities

- Strategic alliances formed with other parties for both R&D and exploitation of an opportunity (see also Section B, 1.3 and Section C, 3.5).
- *Fourth Generation R&D companies*
  - Build on third –generation thinking but Business process focused on innovation rather than technology and product development.
  - Emphasis on the management of knowledge.
  - Technology management represented as the generation of intellectual property.
  - Greater emphasis on discontinuous innovation.
  - R&D works directly with customers, develops prototypes and products via a process of iteration.
  - Organisationally the innovation process uses empowered, multi-disciplinary teams.

For new managers it is a complicated world that they are entering, in which they must be flexible and able to operate in a variety of environments that are constantly changing. In order to carry the fruits of R&D into the marketplace it is essential to work closely and collaboratively with all concerned. There is no point in trying to blame the organisational structure for any problems, it is up to the individuals within a group to get on with the job and resolve any difficulties as they arise using the tools described in this book.

1.2
The Internal Organisation of R&D

A long-standing, shorthand way of describing the spectrum of activities, which make up the totality of research and development, is to use the letters R and D in lower and upper case; e.g. R, R&d, R&D, r&D and D, and to talk about "big R, big R and little d, big R and big D etc".

- R (big R) is pure or fundamental research with no development in mind, most common in research universities or institutes.
- R&d (big R and little d) is product or process research with development only to the level of feasibility studies. Characteristically this work leads to new business ventures.
- R&D (big R and Big D) is an integrated activity from initial research on a new product or process to full-scale development. The type of work done in support of a technologically advanced, existing businesses.
- r&D (little r and big D) is that work that is carried out to establish new products and processes, or to make minor modifications to an existing range. This work is typical of the type of work carried out in many small chemical companies.
- D (big D) involves the incremental improvement of existing products or processes on an ongoing basis. It is a good way of increasing profits, and an essential activity if a product or process is to remain profitable in the long term.

Whilst these shorthand descriptors are useful in clarifying the various roles for research and development activities, they are not in common currency. In most companies R&D is the term that is used to cover the five main areas of basic research, product research, applied research, process development, and technical service [B-7].

Process Development often encompasses the area of process research, and technical service includes applications research. Analytical research is done less and less within companies these days, being left to specialist organisations, and analysis is therefore covered in Section B, 2.1. In bigger companies, the three main areas of R&D, product, process and applied, are often managed separately, but in small companies one group is likely to be responsible for all the activities (see Figure B7). In the context of this book, this difference between small and large companies is irrelevant, as the principles employed in R&D management will be the same in both cases.

1.2.1
R&D Work Groups

In R&D departments, work groups are organised usually in a very similar way to that adopted by the company for its component businesses or functions, and to meet its strategic objectives.

The work groups can therefore be organised by function, i.e. grouped according to disciplines, such as organic chemistry, process engineering, biochemistry, the groups being managed by people with the same technical background. This pure functional/discipline organisational method is now largely out of favour in the chemical industry. However, in a modified form, it is used widely in those companies where R&D is operating a matrix organisation that is built around project teams. As discussed above the functional group is the home for specialists to refresh their competencies whilst waiting allocation to a new project team (see Figure B6).

The equivalents of the Strategic Business Unit within R&D are those groups that are organised around product and application research in particular area, or are allocated to support for a specific manufacturing plant. They will have been given titles, which identify their affiliation, such as the Antiviral Research Group or Multiproduct Plant Development Group. These groups differ from Project Teams by having a continuing role beyond the lifetime of any single project; they have a rolling programme of work. By their very nature they are likely to be multi-disciplinary, for example one associated with a particular plant will be a mixture of process chemists, chemical engineers and analysts. Having people from different disciplines within the same group does present some additional problems to the Manager. This aspect of management is covered in Section A. Because the more sophisticated problems that it faces these days can best be answered by multi-disciplinary or cross-functional groups, this structure is the norm in large sections of the Industry.

1.2.2
Technical Management of Work Groups

Team Leaders play an important role in upholding the reputation of the function.

In R&D, whether a work group is structured according to function, product or business area, it will need to have somebody acting as its leader. In the management line

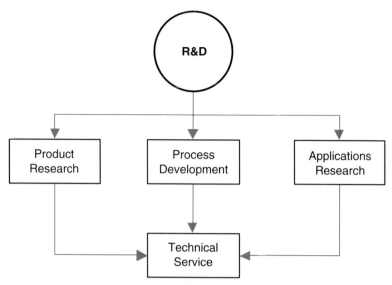

Figure B7. The Internal Organisation of R&D

of the larger R&D departments, a Team or Group Leader, is someone operating at a level below that of departmental manager. These people play a particularly influential role in the building and maintenance of the technical strength and the reputation of the department, especially if they are participating as part of a cross-functional team. They will be heavily involved in the human resource activities, outlined in Section A and in ensuring the safety and well being of the laboratory staff (see later in this Section). However, it is the achievement of the technical objectives of the group and hence the department that is their key responsibility. The position of Group or Team leader is an important step in the career development of people wishing to become senior R&D Managers. In the management of R&D work groups the normal rules of group dynamics apply, especially those relating to interpersonal relationships. The Group Leader has the additional responsibility of ensuring that the technical competence of group members is developed, and in marrying the individual's technical skills to the tasks set for the group, ensuring that any training required is delivered. The group can be viewed as a microcosm of the whole department, being staffed with people having a range of talents and experience. The membership of the group will range from the senior scientists with PhD's, often with many years of experience, through first degree level technical officers with high experimental skills, to relatively inexperienced technicians in the process of gaining the necessary skills.

With the presence of this wide range of technical capabilities there may be problems with communications within the group, especially if it is multi-disciplinary in composition. The R&D Manager will need to ensure that the goals set for the group are discussed with the Group Leader who is in turn responsible for ensuring that all its members understand these goals. Further understanding of and progress towards the goals can be achieved by holding discussions on a one to one basis, and by having regular meetings of the whole team. These actions may seem obvious but Managers often forget them, despite the best quality systems. They are the commonest causes of the technical misunderstandings that occur within groups.

Group Meetings are the occasions when the essential elements of group dynamics come into play. Well-run meetings with close colleagues, involving interplay of technical ideas, are a pleasure for all concerned, especially when the project is going well. However, most of us have been involved in meetings, either as a participant or as a Manager, which have produced the opposite effect and one cannot go back and start the meeting again. Planning for the meeting can help avoid these calamitous occasions.

It is important, from the outset, that all attendees are clear about the nature of the meeting. Is it to be an informal discussion, with free ranging contributions from all participants and open ended outcome, or is it going to be focused and action orientated, with the actions placed on individuals? The atmosphere that exists within the room will be very different for each type of meeting.

The formal Group Meeting has the same basic format and objectives as a Project Meeting. The requirements for the successful running of these types of meetings are as follows.

1. Send out an explicit agenda with a timetable before the Meeting, allowing sufficient time for consideration by the attendees.
2. Clearly outline the subject and object of the Meeting at the start.
3. Direct the discussion. Draw out all opinions, develop group participation but keep the discussion on the subject.
4. Crystallise the discussion. Draw out the pros and cons of each case, and then ensure there is an understanding of the conclusions.
5. Agree an Action Plan. Make sure that it is clear who is responsible for each of the actions and the time for completion.
6. Ensure that a concise record of the meeting is produced, listing all the actions.
7. Set a date for the next meeting if required.

Formal meetings are relatively straightforward to run and, after appropriate experience, success is almost guaranteed. This is not the case with informal meetings where, for no apparent reasons, they can succeed one day and fail the next. This very unpredictability is what makes them a vital source for the cross fertilisation of ideas, of great value in the R&D context. Even though this unpredictability is desirable, it is not a good idea to leave everything to chance. Some guidelines that can help in the running of informal meetings are given below.

- Hold the meetings often, but irregularly, in familiar surroundings. This stops them appearing to be special, and the attendees are more likely to be relaxed.
- Keep the boss away (i.e. never confuse these meetings with the more formal ones, where staff will be updating the senior line manager on developments and feel under stress) and other "outsiders".
- Make sure everybody attends, from the most junior to the longest standing research scientist.
- Get contributions from all staff. Especially the experimentalists and technicians who are the hands, eyes and ears of the group and will have many valuable sug-

gestions to make, if they are encouraged to do so. Additionally, the informal meeting offers an opportunity for the training of junior personnel in the presentation of their work in front of a friendly and receptive audience.

- The quality system will probably require a record of the meeting to be produced. This is against the principle of informal discussions, so make sure it is little more than a record that the meeting took place.

During meetings of multi-disciplinary groups it is essential that no one discipline be seen to be more important than another. There is always some rivalry between scientists of different disciplines, usually of the friendly and joking variety, but it must never be allowed to spill over into antagonism. In a related way the one representative of a discipline in the group, say a microbiologist amongst chemists, should not be viewed as the smartest one in the area, simply because it is the only voice. It is the Manager's job to check that this person is technically competent and performing well, judged by the standards of a larger group of such people (see Section B, 3.4.4). A good way to do this, avoiding any direct or invidious comparisons for the individual concerned is to involve a senior colleague from that discipline in setting the Performance Criteria at the time of Performance Review.

The recognition of a technical achievement within or by the Group needs to be handled carefully. It is always difficult to balance the contribution of the individual and the team. Provided the activities of the group are clear and well understood by everybody, and the meetings described above should have helped to foster this understanding, it should be apparent which people have made major contributions to the success of a project. In these circumstances singling out an individual for praise and reward should not present any problems. In the end the team must feel that a good result was achieved because they worked together as a unit.

1.3
Global R&D

Globalisation of R&D requires an additional set of managerial competencies.

Globalisation is an activity that has been grabbed with both hands by most major companies. It is seen as a way to harness the competitive strengths around the world especially in those countries in the growing economies of the Asia and the Far East, and to foster market penetration in these areas. Globalisation has also been greatly assisted by the availability of rapid international communications, both physical, i.e. transportation and electronic, i.e. telecommunications,

Not surprisingly, R&D has also gone global tapping into the core competencies existing across international boundaries. The input to R&D In these global companies can be achieved in several ways.

- From a team located solely in one geographical location.
- From a truly global team with members from R&D units in more than one country.

- From alliances with other companies in different locations by bringing together their different strengths for a mutually beneficial project.
- From collaborations with universities in other countries by tapping into world class competencies.

Look at the claim from Pfizer to be the company with largest global pharmaceutical R & D organisation [B-38].

> With 2005 actual spending of $7.4 billion in research & development (R&D), Pfizer boasts the industry's largest pharmaceutical R & D organization: Pfizer Global Research and Development.
>
> Pfizer's search for new treatments spans hundreds of research projects across multiple therapeutic areas—more than any other company. Our scientists, clinicians, technicians, and other professionals employ state-of-the-art tools ranging from robotic high-throughput screening (a method pioneered by Pfizer) to sophisticated genomic studies, to deliver a steady stream of innovative new products that enhance human and animal health.
>
> Links with more than 250 partners in academia and industry strengthen our position on the cutting edge of science and biotechnology by providing access to novel R&D tools and to key data on emerging trends.

Another example comes s from Wyeth with what they call their "Global Research Alliances" [B-39].

> At Wyeth, we have forged strategic global research alliances with companies on the cutting edge of technology and innovation. Through our relationships with these companies, we are able to participate in combinatorial chemistry, high throughput robotic screening of compounds, in-licensing of important new products, and genomics initiatives.

Truly trans-national R&D teams bring with them an additional set of managerial and organisational challenges, which include those listed below.
- Team Strategy, setting goals that are understood by all members
- Careful attention to tactics
- Continuous and thorough communications
- An awareness of the cultural identities of each member
  - State and local law
  - Attitudes
  - Time
  - Priorities
- Greater sensitivity
  - Basic culture
  - Interpersonal relations
- Anticipating cultural obstacles
- Managing conflict and competition
- Gaining trust and commitment

In order to help gain the skills required in working in global R&D it is highly likely that a young manager will spend some time in one or more of the company's R&D laboratories in another country.

1.4
Outsourcing R&D

Outsourcing is now an essential component of business strategy.

Outsourcing can be simply defined as the *provision of a service, resource or product from outside the organisation*. At one time the use of outsourcing by R&D managers was tactical, used only when in house provision was stretched or not available, but is now an essential component of the business strategy in most companies [B-8, B-9, B-10].

Why outsource? Innovative companies need to support a vigorous new product pipeline but still want to maintain fixed costs, they cannot hire staff or expand facilities at will. There are three reasons for companies choosing to look at the outsourcing option.

- Businesses are under constant pressure to *control costs,* especially fixed costs, to do more work with fewer people. One answer is to stop any irrelevant support and to consider the outsourcing of non-core, but essential services.
- Everybody wants to get products to market as quickly as possible – the old adage that "time is money" is as true as ever. A shorter *time to market* means that income arrives sooner and the residual patent life left for protected sales is longer. This is particularly true for pharmaceutical companies where the costs incurred during drug development are extremely large. The use of outside agencies to control the peaks and troughs in development is a commonplace activity.
- R&D has become increasingly *technologically complex* and multi-disciplinary. In drug research the advent of technologies such as combinatorial chemistry and high-throughput screening has meant that smaller companies can often provide the expertise required in a more technically advanced and efficient way (see Section B, 2.9). Many projects require multiple skills and few companies, if any, can afford to have all the required technical expertise under their direct control. Outsourcing or joint collaborations are the way round this problem.

What, when and where? Almost anything can be outsourced. Companies can get involved with outsourcing at several different levels.

- Use local organisations or "Yellow Pages" to provide cost effective, commodity services, or services that are only used infrequently by R&D, for example fine chemicals supplies, instrument manufacture and glass blowing.
- Occasional requirements, such as specialised analytical techniques, to be provided on a pay as you go basis.
- Outsource regularly used in-house services that provide no competitive advantage, for example IT and computing, information & library support and routine analytical services, by agreeing an ongoing service contract with a specialist organisation.
- Provision of materials for test. E.g. combinatorial libraries for use in high-throughput screens for new drugs or new materials such as catalysts.
- Custom manufacture of key intermediates and or products, especially to shorten development time.
- All aspects of safety, health and environment, including toxicology.

The development requirements of drug companies include carrying out a whole raft of work on pre-clinical and Phase I, II and III clinical trials (Figure B8). Most of this is ripe for outsourcing and a whole range of contract research organisations (CROs) have sprung able to provide the necessary services for each stage of the development process.

More sophisticated than any of the preceding activities, and one more prone to risk, is the outsourcing of key steps in the discovery phase of R&D. Many managers think that it is very unwise to outsource those technologies that are critical to achieving the business strategy. Outsourcing of only selected research activities and technologies is still the preferred option. Companies prefer to form strategic alliances or collaborate with other companies when essential technological core competencies are missing.

An example from the recent past is provided by DSM, which carries out its research at Geleen in the Netherlands, but also in North America and Asia. In 1999 they reorganised this effort into 25–30 "competence units". In the preceding decade innovations from outside the company had increased from 10% to 50%. At that time the company believed that in future 90% of innovations will start outside. They therefore intended to sustain the innovative process by translating concepts invented elsewhere into innovations at DSM.

"DSM's Venturing & Business Development group continuously scans the horizon for innovative ideas outside the company in which we can invest, enabling us to access and make the best possible use of new technological developments and resources (2004)."

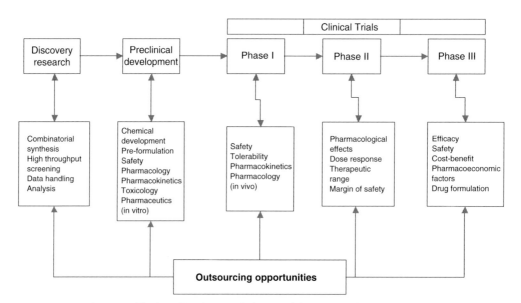

Figure B8.   The Drug Development Pathway and Outsourcing Opportunities

Guidelines

There is no doubt that companies can save fixed and capital costs and deliver projects faster by the intelligent use of outside agencies. However, there are some guidelines to successful outsourcing.

- Commodity services and non-core technologies. These are suitable for outsourcing. And there is a whole consulting industry to provide most, if not all, of those required by the chemical industry.
- Enabling technologies, longer term and or fundamental research. These can all be outsourced but a company needs to retain enough internal intellectual capacity to be intelligent purchasers of the research and technology.
- Critical, core technologies. Need to take great caution, since without core skills a company has little to differentiate it from others in the field.
- Confidentiality and intellectual property ownership. These issues must be addressed and formalised from the outset of any relationship. Do not start if there are any doubts about either.
- Management time. Do not under-estimate the amount of management time that is required to monitor outsourced activities. It has been suggested that if more than 25 % of a manager's ongoing time is spent on an outsourced service it is probably more cost effective to have an in-house resource.
- Suppliers. Make sure they have the right skills, the necessary facilities and can control quality and costs to the desired standards. Preferably choose from prior or existing relationships. Geography, language and culture are important, secondary considerations.
- Planning and project management. These cannot be left to chance. The contract should cover all eventualities and have shared benefits, the supplier being treated as a partner. All should be aware of what is to be delivered and on what time scale.

1.4.1
The One Stop Shop – Super CROs

"One stop shopping" was a late twentieth century phenomenon, whether it referred to a city superstore or a bank which became a financial services centre. By adopting this idea the chemical services industry has targeted the formation of super CROs as the way forward, especially to service drug companies. Super CROs offer a full range of services aimed to reduce drug development times and make the costs more predictable. They operate by forming active partnerships with the lead compound generating research organisations or companies. The potential benefits of working with super CROs are seen to be as follows.

- Able to manage all aspects of an outsourced development programme through a single point of contact
- One legal contract
- One set of Standard operating procedures
- Preferred vendor-supplier relationship
- Bulk purchasing discount

Typically, large CROs compete on the basis of medical and scientific expertise in specific therapeutic areas; the ability to manage large-scale trials on a global basis with strategically located facilities; by providing medical database management capabilities; providing statistical and regulatory services; the proven ability to recruit principal investigators, and patients into studies; and the ability to integrate information technology with systems to improve the efficiency of contract research. The combination of the growing trend by pharmaceutical companies to outsource a wider range of services, and the need to pass products through the testing and regulatory process in a rapid, cost-effective manner, has lead to skyrocketing growth of the CRO market. The CRO market grew from $1 billion in 1992 to more than $8 billion in 2002. CROs enrolled 7 million research subjects in 1992, and 20 million in 2001 [B-40].

There is a distinct possibility that these Super CROs could develop all the capabilities and competencies of a major pharmaceutical company, other then drug discovery area. However, by licensing a potential drug from another source they could move to being drug producers and sellers, thus becoming a challenge to the hands that originally fed them.

1.4.2
The Virtual R&D Company

Tom Peters, the management guru, has been one of the strongest proponents of the "virtual company", arguing that imaginative use of modern information technology renders the corporation redundant.

Virtual companies do not have any technical resource under their direct control and use consultants and other outside agencies to advise on the various stages of the product development. The people managing the company must have very good project management skills to keep the development on track.

Typically a virtual company will license a potential new product from a non-commercial research organisation and then manage the process through to successful launch onto the market using CROs under contract, but often as partners in an alliance. Management is restricted to a Managing Director with a few key personnel, such as a Project Manager, Trial Manager and a Marketing Development Manager, supported by a limited number of non-managerial staff.

Virtual companies are based on the premise that core competencies within chemical and pharmaceutical companies lie in discovery research and not in development. There are many contract research and manufacturing companies who do have the necessary skills for the product development process and the job of the virtual company is to access these skills and then utilise them in an efficient way. The hub in which such a virtual pharmaceutical company operates is illustrated in Figure B9.

Although still in their infancy and still to really prove their worth through the ups and downs of the economic cycle, there is no doubt that virtual companies are here to stay, and will take their place among the variety of companies that make up the speciality chemical and pharmaceutical areas of the industry.

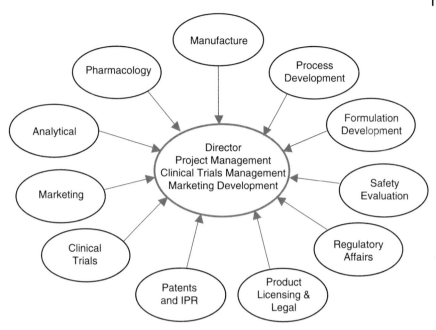

Figure B9. The Virtual Pharmaceutical Company

An interesting development that can be used by companies to answer some of their R&D problems, involves the use of internet sites for forums designed to be virtual knowledge bases, for example InnoCentive. On this website companies can anonymously describe their scientific problems so that they can be solved by outsiders. Over 50,000 scientists have registered with InnoCentive so there is an excellent chance that someone has an answer out there in virtual space. The person who successfully answers the problem gets a not insignificant cash prize from the seeker company [B-41]. Several major companies such as Procter & Gamble and Henkel say they have benefited from the use of InnoCentive [B-42].

2
The Provision of the Appropriate Support

A manager must ensure that the quality of any outsourced services is of the right quality.

In performing their role, scientists working in an R&D area of the Chemical and Pharmaceutical Industries, need the support of a large number of other professional people. Some of these support services are shown in Figure B10. In big companies this support was provided traditionally in house but a majority companies now out-source at least some of these activities to external agencies (see Section B, 1.4). When a service, which previously has been provided internally, is outsourced it is an obvi-ous concern that the support provided may not be of the same quality. To this end it

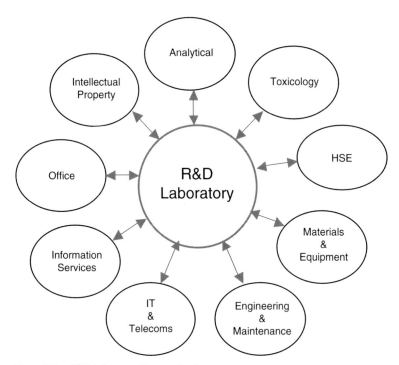

Figure B10. R&D Laboratory Support Services

Research and Development in the Chemical and Pharmaceutical Industry, Third Edition. Peter Bamfield
Copyright © 2006 WILEY-VCH Verlag GmbH & Co. KGaA, Weinheim
ISBN: 3-527-31775-9

is very important that laboratory managers are involved in the drawing up of the contracts and in the assessment of the capabilities of the providers of a service.

2.1
Analytical Services

The ability to analyse the compounds or materials coming out of a chemical laboratory or plant is an absolute essential. It is one of those disciplines that, when properly run and designed to specifically meet the needs of a particular research, development or manufacturing area, can offer a very competitive advantage to the company. Therefore, the wholesale outsourcing of such a facility needs to be thought through very carefully.

A good analytical resource is not cheap. It often has significant ongoing capital requirements for the provision of up to date instruments. It is therefore very important that only those distinctive elements and core competencies are retained within the company. This might well be a particular expertise in a chromatographic method or some advanced methods of mass spectrometry, without which the research could not continue.

Other essential, but non-core analytical services, for instance simple elemental analysis, can easily be outsourced, as they are hardly ever rate determining in R&D, and in addition a rapid turn around in testing is available from competent companies. Outsourcing does not necessarily mean that the work will be done off site. For instance, a visiting contractor can carry out routine environmental analysis of very small quantities of metals in the plant effluent, as is the case with other plant services.

Procedurally the R&D Manager will have to decide, after consulting the relevant staff, what analytical support will be required for each project in the portfolio for the coming period. Following discussions with the analytical services manager, the cost and availability of in house provision can be determined. Analytical services should be able to provide advice on where to get the most economical external provision of non-available analytical support.

The R&D Manager must ensure that the cost for the analytical support is in the budget for each project. Monitoring the cost and the use of analytical services is an essential task in budgetary control as it is easy for an R&D team to run into overspend rapidly. It helps if data is provided to team leaders and chemists on the cost of the methods available in the analytical services group or from the outsourcing agency.

2.2
Intellectual Property

Intellectual property, which arises as the result of work done under R&D programmes, requires to be protected from exploitation by other companies. Arranging this protection requires the services of people who are professionally qualified in this field. The protection usually involves the filing of patent applications covering the inventions made by chemists and other scientists, whether this be new compositions of matter, new formulations, novel chemical processes, devices or anything else which is capable of being protected from exploitation by commercial competitors.

In large companies, a separate department normally exists to carry out this and related important activities, known as Intellectual Property Rights (IPR). In a small company this will not be the case, where an external patent agent or consultant will be used.

Whether there is access to an in house expert or use is made of an outside consultant, a Manager should have a working knowledge of patent law and its operation. Only when this is the case will it be possible to hold meaningful discussions with these experts. The principles and operation of patent law will be dealt with in greater depth in Section C, 2.1.

The IPR people will provide advice on what may be said in public about a company's products and processes, i.e. in papers presented at external symposia by R&D personnel. They will also advise on the use of trade names and marks in scientific papers and other such publications.

When R&D is planned in collaboration with other companies or external agencies, a joint agreement must be drawn up to cover all aspects of the ownership of any intellectual property that may arise. These agreements are written, with the R&D Managers help, by the IP people. This will involve negotiations with their equivalents in the collaborating organisations. An example would be the ownership of IPR in the case of sponsored or joint research with academic institutions.

2.3
Information and Library Services

R&D Managers must ensure that scientists have access to the widest range of high quality information that is affordable.

Paralleling the growth and developments in the storage and transmission of data by electronic means and the ready availability of personal computers, the way in which information is accessed by chemist has undergone a revolutionary change in the last decade.

The days when a library, in a chemical company, was a place filled with serried ranks of books and journals, flanked with large banks of wooden filing cabinets, have long since passed. The books still remain but the filing cabinets have been replaced by gleaming ranks of computer terminals.

The delivery of material via company intranets or from the Internet is now standard practice.

The need for company's to keep any thing more than the books essential for day to day use has passed as delivery from national libraries, e.g. the British Library or from specialist libraries such as that run by the Royal Society of Chemistry, offers a cost effective and rapid alternative. Additionally the move towards publishing on demand will mean that either whole books or selected chapters can be read online and then delivered from publishers as required. Whilst all of the important journals are still produced in hard copy, most are also available in electronic format, readable in house or on a pay for view basis via the Internet. Registration with commercial databases such as Science Direct, allows one to download individual papers and articles for a fee.

The days of the hard copy encyclopaedias of chemical information are certainly numbered. Most of the important ones are now available in CD-ROM format. The data on these CD-ROM's can be conveniently searched, in the home, the office or the laboratory making a trip to the library unnecessary. The highly important databases of international patent information are also available in the same format and also on websites via the Internet. The searching of large databases, e.g. Chemical Abstracts, by inputting chemical structures has lessened the need for an expert to come between the researcher and this information.

Electronic archiving of research reports and test results is now standard practice, and the move towards electronic notebooks is gathering pace. This will mean that the physical storing and the handling of this type of material is becoming less important, especially as documents from earlier periods can now be scanned into an electronic filing system. The role of the library and information services has changed and such services can be economically provided by outside sources. Once again an in house "expert" is required to negotiate contracts and to and monitor their provisions.

All these changes mean that there is less and less need for the information services facility to be in a close geographical position, and most of the activities can be outsourced, if so desired. Since R&D is information hungry, and an essential part of the knowledge based company, it cannot work effectively, without having the ability to search and obtain large volumes of sophisticated information, in as short a time as possible. Whether the service to do this is supplied by an in house resource or externally, should be judged on the basis of quality and cost, not on its necessity, which is a given.

The R&D Manager must ensure that scientists have access to the widest range of high quality information that is affordable. The on-line searching of large databases is not cheap, and sufficient funds need to be in the budget to cover this activity, it is frustrating for scientists not to be able to get the information needed for their work. The people accessing these on line information sources need to be properly trained; otherwise large bills can arise from uncontrolled use of this type of searching. It is a good idea to institute occasional, but regular, checks that the service is being used in a sensible manner. This involves confirming that only data that is really necessary for the project is being collected, that profile searches are cancelled when a project is terminated, as a part of general housekeeping and that the cheapest services are being used.

2.4
IT and Telecommunications

The basic requirement and some indicative applications for IT and telecommunications in the laboratory are:

- *Networks and Communications*
  - Connection into company intranet and external Internet
  - E-mail and efficient communication with colleagues, sharing of files and work group activities. Server for common software. Use of mobile phones both for internal communications and for reporting in when off site. Teleconferencing facility.
- *Databases and Software*
  Standard and chemical Office software. Databases for the storage and retrieval of information generated in the laboratory, with search software. Document management LIMS. Online searching of external databases, catalogues and other information. Electronic Notebooks. Knowledge Management.
- *Modelling*
  Use of dedicated workstations and also networked PCs for 3D-graphics, mathematical modelling and experimental design.

The provision of the services, hardware and software will undoubtedly be outsourced. It is necessary to have a knowledgeable person in house who can assist with day-to-day needs and identify future requirements and to handle negotiations and dealings with the suppliers.

IT in laboratory automation is covered in detail in Section B, 2.9.

2.4.1
R&D Office Technology

Chemists talk to each other by drawing pictures, and when nearly two decades ago software became readily available that allowed this to be done electronically, the office environment for chemists changed in a remarkable manner. Well established commercial office packages allow even the most complicated chemical structures to be drawn easily and incorporated into word processing software; to use electronic notebooks (see Section B, 3.3.2.6); access databases both external and internally on CD-ROMs; integrate into a LIMS; use a registration system; carry out molecular modelling and determine structure activity relationships [B-11].

These easily used packages have transformed the way staff produce reports and other written material and has lead to a minimal requirement for secretarial support. Typing is often at the level of modifying documents etc to ensure common standards in the material produced for electronic or hard copy archival purposes. E-mail and electronic diaries have also had an impact on the secretarial work required by managers. Dealing with external mail, hard copy, e-mail and fax, together with telephone calls and keeping a record of staff holidays and sickness absences are now the main functions of the general office. Making travel arrangements and ensuring the smooth handling of claims for expenses is an activity much appreciated by staff.

R&D generates a large volume of information that has an inherent value, i.e. it will be useful on an on going basis, being part of the company's knowledge base, and therefore must be safely stored for future reference. The material which will be of use to workers in the future includes, experimental data, often in hand-written laboratory notebooks but increasingly in electronic notebooks, test results on materials, including analytical, hazards, toxicity data and application tests, technical and customer visit reports, chemical manufacturing processes and details of the design and layout of manufacturing plants. A large amount of this material is entered directly into computer databases, either at the time the results are generated or during the word processing of reports, or can easily be entered from hard copy using a scanner and support software. Having stored all this valuable data, it is an obvious requirement that scientists be provided with a good means of searching and manipulating this data.

An archive, peculiar to chemistry, is that comprised of the samples of the chemical compounds that are produced during research. Chemists spend many days making these chemicals and confirming their structures by detailed and complex analysis. It is therefore sensible that a sample be retained for future reference. This sample collection serves many purposes, not the least of which is the availability of materials for testing in some future, but as yet unknown, application of interest to the company. The structures of these chemicals, together with relevant test data are recorded in databases. These structure-property databases can be searched using commercially available software, to generate relationships of a quantitative nature (QSAR). More recently the production of combinatorial libraries using automated processing has produced other vast compound databases (see Section B, 2.9.2). Whilst the monetary value of these collections is high, more importantly, they represent the company's efforts in R&D. The samples and associated data need to be as carefully protected and as secure as any other intellectual property.

For the analytical chemist, databases are an essential resource when carrying out their day-to-day activities in product identification. All the major spectroscopic techniques have large databases associated with them, these are used to compare spectra derived from unknown chemicals etc. and so the whole process of product identification is made faster. These are dealt with using a laboratory information management system (LIMS) (see Section B, 2.9.1).

The availability of high powered bench top computers, or the ready access to powerful mainframes via local area networks, coupled with relatively cheap, simple to use software has revolutionised how computer aided design is carried out. This is no longer within the confines of an expert group inside the larger companies but can be carried out even in modest sized companies in the chemical industry. This has been particularly beneficial to process engineers when designing chemical plants and for chemists working at the boundaries of biology and chemistry in the biological and medicinal chemistry sectors of the industry.

2.5
Health, Safety and Environment

Looking after the health and safety of people in the workplace is both a legal and an ethical requirement in most countries throughout the world. It is therefore a given that it is the responsibility of the R&D Manager to ensure that the provisions of the various pieces of legislation are met for all staff within the building. The day-to-day administration will usually be delegated to a responsible member of staff. Details of the various aspects of HSE policy are covered in Section B, 3.2.

2.6
Toxicology

The testing of the toxicology of a new chemical is required at some stage in the development of new products and is a very expensive procedure. The decision to go ahead will be made at a business level based on likely sales and profitability of the new product. A laboratory specialising in toxicology will carry out this work. Professional advice on the likely toxicity of materials, based on analogies, can be obtained from consultants at a very early stage, thus possibly avoiding expensive development work, which terminates on the discovery of toxicological problems (see also Section B, 3.3 and Section D, 2.2.2.1).

2.7
Chemicals and Equipment Supplies

The chemicals required for use in the laboratory fall into the following basic categories.

- Fine Chemicals, from suppliers
- Bulk Chemicals, including solvents
- Media, e.g. gels, resins
- Samples from the Manufacturing Plant
- Samples produced in the laboratory

The first three of these categories are readily available from local suppliers, whose catalogues are available in a variety of formats, hard copy, on CD-ROM or on-line, including on-line ordering. Therefore only minimal buffer stocks of these supplies need to be held in the laboratory, and to operate a "just in time" system. Fine chemicals in particular are expensive and overstocking is an unnecessary financial burden on the R&D budget. From the practical point of view, the control and monitoring of expenditure by laboratory staff, is most easily done if an individual is identified to carry out the ordering on behalf of the group. In addition, an identified individual is more likely to strike up a good relationship and understanding with the suppliers.

Manufacturing samples, by their very nature are often large and bulky and the provision of adequate and safe storage space in a convenient location can present problems that are not associated with fine chemicals.

The storage and retrieval of laboratory samples, on a longer-term basis, has already been discussed in Section B, 2.4.1. However, many of these samples will be in regular use and therefore will require storage space either in or close to the laboratory. Systems have been devised to assist in locating these locally stored materials and so making them more readily available to the occupants of other laboratories in the building. The simplest of these systems uses a bar coding system. Each bottle or vial containing a chemical is given a bar code designated to its contents. The location of the chemical is then entered into the database on the network using a bar code reader. The bar code reader is then used to record into the computer when the chemical is moved to another location. A user can check the location of a particular chemical from the computer database. These systems only work when everybody follows the procedures. Properly used, these methods can rapidly pay back the investment in the savings made from the reduced purchases of fine chemicals. All chemicals need to be handled in a safe manner and this is discussed in Section B, 3.2. In order to comply with the regulatory authorities (see Section B, 3.3), it has become imperative to track the movement of samples around the laboratory and inventory management systems to do this are available commercially.

The equipment used in R&D laboratories covers a wide range, from simple, cheap glassware to sophisticated and expensive automated synthesis rigs, analytical machines, testing equipment and pilot plant items. It is one of the jobs of an R&D Manager to determine the equipment required by the group and to plan for its acquisition. All laboratory equipment has a finite lifetime and as well as a process of planned maintenance, a plan for its replacement is required. It is usually better to leave the purchase and even the choice of supplier, to a person charged with a purchasing and supply responsibility for the department or company (see also Section B, 3.1.1)

2.8
Engineering and Buildings Maintenance

This basically covers those activities that are required to manage and maintain the physical fabric and the supply of utilities to the building, namely.

- Keeping the building and infrastructure sound, and in clean, safe working order.
- Control of energy supply; electricity, coal, oil and gas.
- Water, effluent and other waste management.

Whilst most of these activities can or will be carried out by outside contractors, there must be a member of staff, or even a small team, responsible for ensuring that the contractors meet their obligations of their contract with the company.

2.9
Laboratory Automation

Developing a strategy for automation in the laboratory involves making sure that it meets the needs and objectives of both the laboratory and the company.

For many years automation in R&D was driven by the need for the analytical laboratory to increase its productivity, i.e. having a faster turn around time in sample analysis, improving quality and reliability, freeing staff for more productive or creative work, increasing safety by having less manual handling of potentially dangerous chemicals and by lowering the consumption of a sample material in a test procedure [B-12].

However, all this changed with the development by pharmaceutical companies of high throughput screening methodologies for the testing of new drugs during the late 1980s. These screens required very large numbers of samples to keep them busy, far more than even a large group of synthetic chemists could muster. New methods of producing the large number of samples were needed and this lead to the development of automated synthetic methods, called combinatorial chemistry and the production of so-called libraries of compounds The original methods were for libraries of peptides, based on the Merrifield solid phase synthesis first devised in the 1960s, but this was extended to small molecule organic chemistry in the 1990s [B-13].

Only applied originally in the pharmaceutical industry, automated synthetic methods have spread quickly into the research for new agrochemicals, other speciality chemicals, catalysts and new materials. It is also being used increasingly in process development laboratories coupled with statistical experimental design.

Automation requirements will not necessarily be the same for an analytical and a synthesis laboratory or from different laboratories in the same organisation. The benefits from automation need to be considered on a case-by-case basis.

Managers, when developing a strategy for automation in the laboratory, should make sure that it meets the needs and objectives of both the laboratory and the company. To this end the following actions are recommended.

- Identify where automation can improve the productivity of a laboratory environment and what contribution it can make to the overall aims of the organisation.
- Show there is consistency with the business objectives of the laboratory and those of the organisation.
- Design a strategy that meets the user and the laboratory's own needs.
- Ensure there is integration of the laboratory with the organisation or at least the laboratory with users of the information.
- Show there is consistency with the remaining laboratory procedures and if some procedures need changing that these will be done in a quality controlled manner.
- Only make those changes which are realistic and cost effective.
- Ensure consistency with the IT systems (i.e. hardware, software and communications standards) of the organisation.
- Build in a review of and revise the strategic plan at appropriate intervals and revise as necessary.

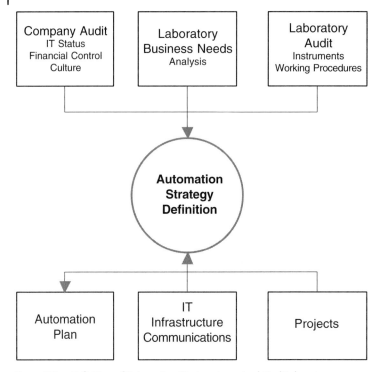

Figure B11. Definition of Automation Strategy in an Analytical Laboratory

The process for implementing an automation strategy is illustrated shown in Figure B11 [B-12].

2.9.1
The Analytical Laboratory

Analysis by its very nature requires the repetition of standardised techniques and is an ideal area for automation. Automation in the analytical laboratory comprises four groups of activities.

1. Instrument automation. Designed to speed up the tasks of sample preparation and analysis.
    - Autosamplers
    - Autoanalysers
    - Robots
2. Communications. The means by which information is passed from equipment on the laboratory bench to the analyst and then on to the user/decision maker.
    - Networks
    - Multimedia
    - Client servers

3. Data to information conversion. Converts the data from analytical instruments into useful information.
   - Chromatography data systems software
   - Chemometric techniques
4. Information Management. Stores the data as it is generated to be used for onward delivery of information and hence the creation of knowledge in the user.
   - Databases
   - LIMS
   - Neural networks
   - Knowledge management

In all automated analytical laboratories a *laboratory information management system* (LIMS) will be used and this has two key roles [B-12].

1. *In the laboratory.* Performs the function of an information generator.
2. *In the organisation.* The means by which information is presented to the users.

LIMS is therefore an extremely important component of analytical automation, providing laboratories with the means to automate the process of information creation and then a means of distributing the information to managers and clients, so that the appropriate decisions can be made.

Many commercial LIMS system are available and the choice is governed by how well any system meets the needs of the automation strategy of the laboratory and company. The introduction of any LIMS system will be handled by a project team made up of laboratory management, IT professionals and representatives from the user community.

2.9.2
The Synthesis Laboratory

There are basically five elements to the automation currently applied in the synthesis of molecules and materials in the research and development laboratory:

1. Combinatorial Synthesis
2. High Throughput Screening
3. High Throughput Experimentation
4. Process Development
5. Microreactors

Combinatorial chemistry is the production of libraries of compounds that represent permutations of a set of chemical variables. These variables include the nature of the substituent in a particular molecule, both in type and size, changes in the components in a mixture of materials, e.g. in ceramics and changes in process parameters, e.g. temperature, pH etc. Chemical libraries are usually created by one of two methods: "split and mix" or "parallel synthesis". Split and mix synthesis is used to produce small quantities of a relatively large number of compounds and requires assays to be performed on pools of compounds. Parallel synthesis is used to produce libraries of single compounds and in process development. Several, detailed accounts of combinatorial chemistry are available [B-13].

High Throughput screening is the rapid assessment of a specific property for each of a large number of samples. The format for screening may be parallel, in which one or more samples are assessed simultaneously, or sequentially.

High Throughput Experimentation grew out of combinatorial chemistry and is defined as: "the rapid completion of one or more experimental stages in a concerted and integrated fashion [B-14]". It comprises the integration of the synthesis, processing, and characterisation, testing and screening stages with the design and modelling stages in a suitable informatics environment. A simple schematic of the cyclical/iterative process is shown in Figure B12.

The design stage is where decisions are made on which particular molecular structure will form the basis for the subsequent library, what structural variations are to be made on this basic structure, what are the synthetic steps and what analytical methods and test procedures will be used to evaluate these structural variations. The experiments to be carried benefit significantly by being designed using statistical experimental design methodology. Split and mix and parallel synthesis, especially the latter for small molecules and projects requiring sizeable samples for evaluation,

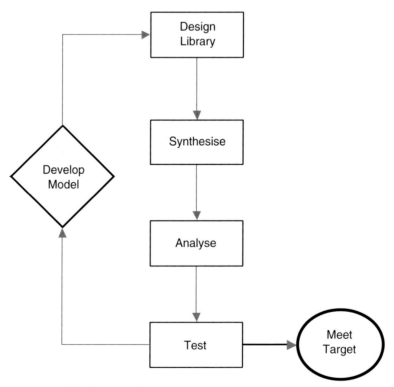

Figure B12. High Throughput Experimentation

are then used to build up the compound library, isolation being achieved by physical separation e.g. centrifuging of solids, distillation of liquids or by chromatographic techniques.

Rapid analysis of the materials in the compound library is essential if the design process is to be an iterative process and integration of systems into the automated analytical library, as described above, is the ideal. Automatic recycling of this information into the design process via a LIMS is desirable. Preliminary screening against single target or specific test is achieved readily by high throughput screening methods. However, for many speciality chemicals the testing step is likely to be rate limiting, as it is unlikely that there will be a simple yes or no answer. The objective of the work is to exit with a product, which meets the initial research target and can be further refined into a development product. All the results from the synthesis, analysis and testing stages are used to further develop the model and as a basis for going through another cycle. To this end this end quantitative structure property/ activity (QSPR/QSAR) packages can be used to help predict the properties of new molecules.

Automated workstations are ideally suited for examining the various reaction stages and unit operations involved in the process of making a chemical. Several off the peg commercial systems are now available.

2.9.2.1 Synthetic Automation Specification

Although many excellent commercial systems exist, it is still often necessary to customise these to a particular users needs. It is all too easy for both managers and users to over specify their requirements, making the system more expensive than it needs to be. Consequently a few important considerations taken at the outset of a project can lead to rapid success using the currently available technology.

- This should usually combine proven techniques and reliable hardware, with flexibility for future developments.
- The supplier should show a good track record with financial stability to support the customer for the expected lifetime of the equipment.

Another important consideration is the automation company personnel:

- Automation companies should have attracted scientists who are able to understand the true needs of their customers and avoid impractical projects.
- A clear specification for any automation system is needed and the subsequent project schedule, the overall impact within the organisation should be analysed.

The introduction of automation may not be seen as successful, if the automation system moves a "bottleneck" to a new location within the organisation. Relevant and enthusiastic users are always required to work with the automation company, ensuring the success of the project and a rapid return on investment.

- Analyse requirements, not wishes
- Have clear criteria for success
- Employ relevant and enthusiastic users
- Understand the implications and impact

Before the final decision is taken on the direction of an automation project, the following questions should be asked in the laboratory:

- Is there a system available that performs to meet all our requirements?
- Do we need to demand a system that meets all our requirements?
- Should we move ahead with today's offerings, even with the shortcomings?

Laboratory managers will be in a much better position to have meaningful negotiations with equipment suppliers after having canvassed the opinion of all users.

2.9.2.2 System Integration and Throughput

Unless laboratory staff are competent with any new technology its introduction into the laboratory will present difficulties. Many bottlenecks need to be removed if maximum benefits are to be obtained from the technology, and efficient data handling and experimental design are essential.

Developing an experimental strategy that gets the most from automation and parallel approaches to experimentation requires somewhat of a different perspective from traditional ways of working, and training or experience is therefore essential. High throughput experimentation in chemicals R&D projects may require a change in overall approach to research. The various elements of the automated technology platform are much more expensive than the manual approach and significant new developments of screening methods might also be necessary for a given chemical project.

However, as in many other R&D areas, the most critical to the success of an overall program is the level of staff competence in the technology. In the largest chemical and pharmaceutical companies making a large investment in technology platform, development and expertise is done internally. For others, the preferred option is to outsource the components of those research programs entailing high throughput experimentation elements.

The overall throughput, as within any chain or cycle, is constrained by the slowest steps and it is essential to focus on and eliminate these successively. Overcoming key bottlenecks to enable a fully automated system requires considerable work on the chemical development of the various steps that are often done off line, such as isolation and washing. This can be overcome partially by a systematic approach in the design stage where the conditions for each of the reaction stages must be defined such that yields are optimised across the range of substrates represented by the library.

Computation plays a critical role in three areas:

- Experimental Design
- Data Handling
- QSAR/QSPR

Using the techniques of Statistical Experimental Design, at the design stage has become key, whether this is for compound libraries or for the parameters in the process development of a product. A large and complex compound library generates a great deal of data to be handled and an efficient information management system is just as essential as in the analytical laboratory. Library design will involve certain parameters in the molecules that can be measured, enabling these to be used to produce descriptors for input to QSAR/QSPR packages. The correlations between composition and properties of the molecules are then used in further refinement of the molecular or product design to meet specific research targets.

2.9.2.3 Lessons for Management

After more than a decade of development there are several key messages for the management of automated systems in the synthesis laboratory. An interdisciplinary approach is needed:

- Applications in speciality chemicals are diverse.
- Business value drivers are important.
- Business models differ across the industry.
- Chance discoveries are more likely.

The development and application of high throughput experimentation for a given application requires an interdisciplinary mix of Chemistry, Physics, Engineering, and Computation in order to develop the chemistry, software and mechanical modules.

Drug discovery programs are usually designed to identify an organic molecule in the 200–600 molecular weight range that has the optimal profile of properties for a given indication. Although this profile ultimately includes absorption, distribution, metabolism, excretion and toxicological properties, the primary screening vehicle is the binding of the molecule to a protein target. Therefore automated machinery for testing can thus be applied directly in many different drug discovery programs. However, in many other areas, such as specialty chemicals, the testing of samples involves the measurement of a range of diverse properties, e.g. smell in fragrances, corrosion inhibition in metal protection, colour in the solid state for dyes and pigments, permeability through a polymer for additives, viscosity in paints and inks. Because successful drugs generate sales of >$1 billion per year, major investment in early research to discover potential new drugs is easily justified, but in specialty chemicals programmes, where the returns on a new product are much lower, there is a greater need to analyse carefully whether the value of a significant advance warrants the relatively high initial investment in automation. In these cases outsourcing may be the best option, certainly in the evaluation phase of a project.

In the early days of automated synthesis there was a fear amongst researchers that it would reduce the chances of serendipity. However, experience has now shown that, when used in sensible way, the, automation of the design, synthesis and test cycle can increase significantly the odds of chance discoveries of new compound that have attractive properties to the company in unforeseen ways.

2.9.2.4 Microscale Experimentation

Micro reactors and devices that can process and analyse very, very small amounts of chemicals is an area that is rapidly advancing, with applications in both the analytical and synthetic laboratories and also in full-scale manufacture [B-15, B-16]. This area is also known by the jargon term, lab-on-a-chip, as many of the devices are sub-millimetre scale, integrated chemical systems built by the microfabrication techniques developed in the electronics industry.

This reduction in size and the integration with multiple functions offer capabilities beyond the conventional macro reactors, namely:

- Adding new functionality
- Potentially lower cost due to the mass production techniques common in the chip industry
- Lower maintenance and operation costs
- Lower power consumption
- Safer operation

The miniaturisation of analytical devices has obvious uses in high throughput screening in drug discovery and the developments are being driven along by collaboration between instrument manufacturers and the large pharmaceutical companies. The term "miniaturised total analysis system" or "μTAS" is also used in the analytical field.

In the first decade of this new century, micro reactors are forecast to take over many of the function currently carried out in the laboratory and make their mark on the manufacturing plant [B-43, B-44]. To date the main areas of development have been:

- Analytical systems for DNA sequencing
- Systems for high throughput screening
- Analytical Systems in defence uses against chemical and biological weapons
- Devices for point of care clinical analyses
- Microreactors for hazardous chemical reactions
  "DuPont have described a microreactor for carrying out hazardous reactions. This is fabricated from layers of wafer-like disks with precise interior channels formed on the disk surfaces to contain the reactions. Each microreactor is an integrated system designed to perform a specific process including mixing, heat exchange, catalysis, reaction, photoreaction, electrochemical, separation and analysis/control of reactions involving gases, liquids, as well as multiphases. The channels within these layers measuring a mere 10 to 5000 micrometers across and connected to the inlet and outlet ports are formed by any number of techniques depending on the material used: chemical etching, electrochemical machining, laser machining, electroforming, selective plating, chemical vapour deposition, photoforming, moulding, casting and stamping. The size and versatility of these microreactors make them ideal for lab bench processes as well as volume production in which dozens, even hundreds, of devices can be operated in parallel to produce potentially millions of pounds of chemicals a year" [B-45]
- Continuous processing in the production of fine chemicals
  In 2004 Clariant set up a new centre for the development of microreactor technology in Frankfurt, Germany, called the Clariant Competence Center for MicroReaction Technology. Its facilities include two microreactor pilot plants and a dedicated laboratory. Clariant says that by using parallel microreactor modules makes it possible to turn out kilogram or even ton quantities of a product [B-46],

With the use of microreactors rapidly increasing, the laboratory manager, in dealing with what will certainly be multi-disciplinary teams working at the forefront of technology, will have to employ the interpersonal skills described elsewhere in this book.

3

A Financially Sound, Healthy, Safe and Quality Environment

The working environment must be financially sound, healthy, safe and organised in a quality manner.

The working environment in R&D, like any other area within a chemical or pharmaceutical company, is required to be financially sound, to be healthy and safe and to be organised in a quality way. However, there are many special features of R&D that are not found in other functions and these are the ones that will be considered in this part.

3.1
Financial Control

The financial control of R&D group is much simpler than in some of the other functions where chemists may find themselves in a managerial role, for instance in production and manufacturing, or sales and marketing. This is because there are few day-to-day financial transactions, except in charging out services like analytical, and there is only a limited stock of materials and equipment to control. There are three main elements in the financial control of an R&D group.

1. The setting of forward plans, usually annual ones
2. The construction of the budget, again usually annually
3. The ongoing control of costs within the budget

The setting of forward plans and the construction of a budget for that period are symbiotic, the one cannot exist without the other. The size of the annual plan agreed with the executive or managers of individual business units will, in the end, govern the budget limit for R&D, However, the cost of each component of this plan, and hence whether individual business units can afford to do the desired R&D work, will only be known when all the elements within the overall cost for the R&D group, the budget figures have been worked out. Ideally, plans and budgets should be drawn up at the same time if surprises for both parties are to be avoided.

Research and Development in the Chemical and Pharmaceutical Industry, Third Edition. Peter Bamfield
Copyright © 2006 WILEY-VCH Verlag GmbH & Co. KGaA, Weinheim
ISBN: 3-527-31775-9

3.1.1
Budgets

In the first instance the construction of the budget for the R&D group is done on the basis of the existing complement of staff. The problem of having an oversized budget is considered in Section B, 3.1.2.1. There are several elements that form the total budget, but within R&D personnel cost always dominate.

There are many ways of drawing up a cost budget and every company will have their own formalised system, which an R&D Manager will have to follow and understand. However, it is unlikely that these systems will vary markedly from one company to another even across geographical boundaries. Typical components of a budget setting system, which a new Manager is likely to come across, are described below.

Each of the individual elements of a budget requires careful attention by the Manager, as an error at this stage can later lead to an overspend, something which is never liked by business managers or accountants. The largest single element in any R&D budget is, as mentioned earlier, the cost of personnel, often as high as 80% of the total, excluding site or company overheads. When calculating personnel costs, it is easy to underestimate the impact of events which occur during the year on the eventual outcome of R&D expenditure. These include additional costs arising from the promotion of staff, increases from performance related pay figures and items beyond a Manager's control, such as a general cost of living rises for all staff. There is also the effect of national or international inflation on the costs of other items, especially consumables and minor capital items. The cost for the training needs of staff, arising from the performance management process, need to be estimated. Some projects may require a large number of visits, e.g. to customer's premises, to plants and facilities in remote locations, and the travel budget needs to be set following an estimate of these activities. The make up of a typical cost budget is shown below.

Cost Budget
Personnel Costs
   Salaries
   Associated Payroll Costs
Total Personnel Costs   A
Consumables
   Chemicals
   Equipment
   Stationery
Minor Capital Items
Engineering
IT Charges & Telecommunications
Information Services
Training
Travel & Entertaining
Department Overheads
Total Other Costs   B

Total controllable Cost (A + B)	X
Site & HQ Overheads (non-controllable)	Y
Total Overheaded Costs	T

The figure X is often described as a controllable cost, one that is within the control of a Manager. Depending on the system employed by the company, there will also be associated costs coming from other essential HQ functions, e.g. Human Resources, and a site overhead for servicing the building. Therefore X will be inflated by a significant figure, Y. An R&D Manager may well be involved in negotiating the service level contracts for the components of Y. It is on these occasions that any training received in negotiating skills will be put to the test.

The total cost for the R&D group needs to be allocated, equitably, to each of the projects. This can be done in several ways, but for R&D this is best done on the basis of the number of people working on each of the projects. There are two ways of doing this; the global or the specific method.

1. *The global method*

   The total overheaded costs T are divided by the total number of R&D staff in the laboratory deriving a cost per unit of N. The cost of an individual project is then calculated using this unit cost. For instance, if a project requires n staff to be working on it then the cost of the project is n x N. This method does mean that a technician costs as much as a senior research worker and is not suitable for use in small R&D groups where the actual project cost would be distorted. However, in a large group, provided there is no marked imbalance in the levels of staff on each of the projects, the costs even themselves out over several projects and the system is very easy to administer and monitor. It can be made more accurate by applying the same process to different bands of staff, such as technician, technical officer, and senior technical officer etc., giving N1, N2, N3 etc.

2. *The specific method*

   The salary and associated payroll costs for each individual member of the project are added together, making sure that an allowance is made for changes to these numbers, e.g. performance related pay rises or bonuses, during the reporting period, to give a total figure of C. The non-salary related costs, the sum B and Y, is then divided by the number of personnel on the project (n) to give D. The total cost for the project is then the sum of C and D. This method gives a more accurate cost for each project, but requires a greater attention to detail opposite movement or exchange of personnel during the lifetime of the project, which will affect the larger salary related element of the cost.

Whichever method is employed, the figures obtained are applied to each of the projects in the annual plan. The resultant cost for each project is then communicated to the business for inclusion in their budget.

3.1.2
Plans

The methodology used to decide which projects should make up the annual plan, the targets for R&D, is described in Section D. The annual plans for technical service or analytical support will normally be drawn up to match the requirements of the projects in the R&D plan, with additional items covering business and manufacturing support.

The costs associated with these annual plans are twofold; the costs of the personnel involved and the capital requirements.

3.1.2.1 Personnel Costs

Staff are allocated to each of the projects, in the first instance, on the basis of the number and level required to achieve the projects objectives in the desired time. Using the cost per person, N or N1, N2, N3 etc. as described in Section B, 3.1.1, it is possible to derive the cost of each project.

> As an example, let us consider the case of a group within R&D carrying out work on behalf of the Surfactants Business Unit. Three projects have been identified by the business as being crucial for the next year. The R&D manager has made an estimate of the staff required for each of the projects. The cost for each of the projects is then calculated and hence the total cost to the Business Unit. A simplified version of the annual plan, showing only the financial elements and using N as the unit cost per person, is given in Table B3.

Table B3   R&D for Surfactants Business Unit

Project	Number of People Equivalents	Cost (Unit=N)
A	3	3N
B	1	1N
C	5	5N
Total		9N

It is at this point, after discussion with colleagues in the business unit, that any financial difficulties will be identified. If the costs are too high, either in total or for an individual project, priorities will need to be agreed and less effort put on the projects of lowest priority. It may be the case that the figures come as a pleasant surprise to the Business Manager, who is then willing to see more people allocated to the projects in order to get results in a shorter time. Although this is good news for an R&D Manager, it might not easily be achieved, the staff with the necessary skills may not be available and there could be a need to either retrain existing staff or recruit extra people.

3.1.2.2 Capital Costs

Capital investment in the R&D facility is a constant requirement, especially in the instrumentally intensive automated areas. The analytical laboratory is one of those areas where instruments are constantly being updated and new ones introduced. The semi-technical plant is another area where significant costs can be incurred, and these will need to be allocated to a project. These capital requirements must be under constant review and considered during the planning phase of any project. There is little point in starting a project if an essential piece of equipment is not available or cannot be hired. An early warning should be given to the Business Manger if capital is going to be required for a project, so that a figure is put into the business plans for capital expenditure.

3.1.3
Cost Control and Monitoring

Controllable costs, i.e. those that an R&D Manager will be expected to control have been described in Section B, 3.1.1. In large R&D units, expenditure on consumable items, local travel and training costs can be delegated to the more senior scientists. It is still surprising to find that in many companies, even quite senior people are not allowed to sanction small sums of expenditure. A person who is successfully running a home, with all the financial dealing and control involved, when in work is not allowed to sanction the purchase of a new refrigerator for the laboratory, without higher authority. When there are computer aided ordering systems in place that allow checks and balances to be employed, this attitude to minor expenditure seems archaic.

The control of the costs allocated to each project is the next level of cost control for the R&D Manager. Monthly auditing is normally employed by companies, with the derivation of the necessary data easily done if the cost per unit system, described in Section B, 3.1.2.1, is used. It should be based on record of time actually spent on the project not on a notional figure.

Let us again consider the example of the three projects being undertaken for the Surfactants Business Unit, given in Table B3. Four months have passed and the expenditure has been allocated as shown in Table B4. Each project can be compared with the expected spend in the annual plan. Project A is under budget, but this is not a problem since it is known that a higher rate of spend is due in the second half of the year. Project B is in line with planned expenditure. Project C

Table B4 Project Cost Control; Surfactants Business Unit, Four Monthly Return

Project	Number of People Units	Four Monthly Costs (Actual)	Four Monthly Costs (Budget)
Project A	3	0.66N	1.0N
Project B	1	0.33N	0.33N
Project C	5	2.00N	1.67N
Totals	9	2.99N	3.00N

is ahead of plan, but once again this is all right, more people were put on the project in the early part of the year but will move elsewhere in the later period. As can be seen from Table B4, the total expenditure is very close to that the expected in the budget. If control were operating at this level, the variation in the individual components of the budget would be missed. Therefore, control of the costs in R&D should be at the individual project level and not at the group total for the Business Unit.

3.2
Health and Safety

All members of staff are have a general responsibility in health and safety legislation but R&D management have specific responsibilities.

Legislation covering the health and safety of people in the workplace is operational in most countries throughout the world. Of particular interest, to those working in the chemical industry, is the legislation that is concerned with the protection of people who will be exposed to chemicals. It is therefore an essential requirement that an R&D Manager, having a responsibility for people who are working in laboratories or on a chemical plant, be familiar with the legislative background within which the company and its employees has to work.

3.2.1
Legal Background

For the countries of the European Union (EU), the basic law is enshrined in the Single European Act (SEA) of 1986, through articles 100a and 118a. These articles cover the duty for each member country to legislate on health and safety. Many EU directives have been passed since that date and these have subsequently been covered in the legislation of the individual countries. For companies who are operating in a particular country within the EU, it is the law of that individual country which is applicable.

The relative legislation in the individual countries of the EU include the following: Order on Dangerous Substances in the Workplace (Germany), the Arbo Act (Netherlands), the General Regulations of the Protection Labour RGPT (Belgium), the Labour Code (France) and the Control of Substance Hazardous to Health COSHH (UK). Codes of practices have been drafted to assist in the management of the HSE processes. For instance, in the UK when these have been officially sanctioned by the Secretary of State they are known as Approved Codes of Practice (ACOP)

The provisions of the COSHH act will be given in outline, being typical of those in force in other countries [B-17, B-18, B-19].

- Collecting and evaluating information on the chemicals likely to occur in the laboratory.
- Assessing the risks to health in the various work activities performed by employees.

- Deciding on appropriate measures to be applied to control or prevent those risks.
- Ensuring proper maintenance of the control measures.
- Monitoring the workplace atmosphere to ensure that containments are kept below harmful levels.
- Carrying out appropriate health surveillance.
- Providing appropriate information, instruction and training of workers.
- Keeping careful records of the above activities.

In the United States the basis of the relative legislation covering chemicals in the workplace is the Occupational Safety and Health Act (OSHAct) which is enacted via the Occupational Health and Safety Administration (OSHA). The objectives of this body are as listed below [B-17].

- To encourage the reduction of workplace hazards
- To provide for research in occupational safety and health
- To establish "separate but dependent responsibilities and rights" for employers and employees for the achievement of better safety and health conditions
- To maintain a reporting and record-keeping system
- To establish training programmes
- To develop mandatory job safety and health standards
- To provide for the development, analysis, evaluation and approval of state occupational safety and health programmes

By comparing the COSHH and OSHA approach it can be seen that there is great commonality. Both are based on a common sense attitude to safety in the workplace, especially the laboratory and chemical plant.

3.2.2
Management of Health and Safety

All chemical companies will have a person, or group of people, depending on their size, directly responsible for health, safety and environment (HSE) and for regulatory affairs. All model HSE management systems consist of a Plan, Do, Check and Review cycle as illustrated in Figure B13 [B-20]. The Check and Review stages enable any necessary corrective actions to be taken.

Whilst company-HSE staff will provide professional guidance to an R&D Manager and all other staff in such matters, it is the duty of everybody in the workplace to behave responsibly in matters of health and safety. This is the ethical dimension of HSE, which underpins the Code of Conduct for members of professional societies such as The Royal Society of Chemistry. Whilst all members of staff are subject to a general responsibility in health and safety legislation, R&D management have specific responsibilities, which should be covered in the company's safety manual or procedures. It is the R&D Manager's duty to read these procedures and to be conversant with the provisions covering both direct responsibilities and those of the laboratory staff. Every worker in the laboratory must be trained in the provisions of the legislation and how these apply to their particular job.

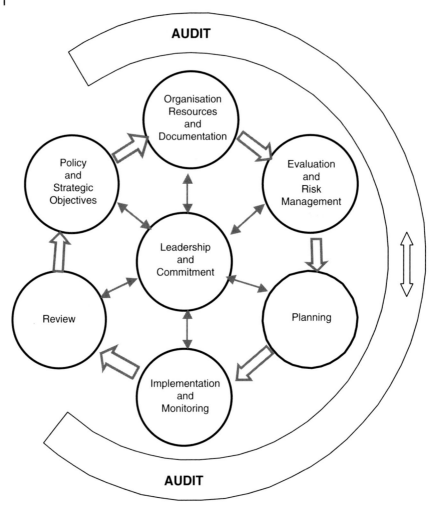

Figure B13. The HSE Management System

3.2.2.1 Risk Assessment

The chemical laboratory is a potentially hazardous place. With such a potential for danger it is remarkable how few really serious accidents occur. This must be due to the rigorous training scientists receive, especially chemists, building an understanding of the extent and nature of chemical hazards. Under the provisions of the legislation, Managers must ensure that all new recruits are trained and fully familiar with the safety procedures operating in the laboratories. It is also wise to have in place a system whereby new recruits, during their initial period of familiarisation with the laboratory, are in the hands of an experienced mentor.

Potentially, the most lethal events can occur in those laboratories where chemicals and biologically active materials are being handled. As mentioned above, in the UK the Control of Substance Hazardous to Health (COSHH) covers this and similar legislation is operational elsewhere in the world [B-18, B-19]. Under this legislation all materials must be labelled according to their toxicological hazard rating, and for this to be noted in all laboratory notebooks when they are handled. Hazards in the chemical laboratory are not only chemical and biological but are also physical and ergonomic, hence the need to consider the hazards associated with all unit operations and other major activities in the laboratory.

There are therefore specific requirements placed on managers to ensure that risk assessments are carried out in the laboratory on all likely hazards [B-21]. Risk assessments are designed to make an estimate of the chance of harm or injury arising from likely exposure to a hazard and then making a decision on what the response should be to reduce the risk to as low as is reasonably practicable. A competent person or persons, depending on the complexity of the situation, must carry out the risk assessment. In addition the systems operating in the R&D group must be able to stand the rigour of an audit by external experts. The procedure for a risk assessment exercise requires the competent person to carry out the following steps:

- Identify the hazards
- Estimate the likely exposure to the hazard
- Evaluate the risk
- Ensure that there is compliance with specific regulations.
- Consider what the consequence will be from overexposure to the hazard.
- Decide whether control is reasonably practical or needs further action.
- Document existing or required control measures
- Review the assessment on future occasions to check action has been taken, if required

Having identified a hazard there is a hierarchy in the controls that can be put in place to deal with the hazards. Starting with the preferred option of hazard elimination the list is as follows:

- Elimination of the hazard – stop the activity
- Substitute the hazard – find a safer replacement
- Segregate the hazard – remove to an isolated area
- Engineering controls
- Work systems – permits to work
- Training and information
- Personal protection

3.2.2.2 Performance Standards and Indicators

Performance standards need to be set by management at such a level that the business is able to function whilst minimising hazards and risks to people and the environment at every stage [B-22]. In the laboratory these performance standards need to cover three operational aspects.

1. *Control of Inputs.* The objective is to eliminate hazards and minimise risk associated with equipment, chemicals and materials, information and the people who

come into the laboratory. For example HSE performance standards set with contractors who supply chemicals or materials.

2. *Process Control.* The objective is to eliminate or minimise risks to people whilst carrying out their daily tasks in their working environment. For example by having performance standards for all laboratory operations and the maintenance of equipment.

3. *Control of Outputs.* The objective is to control risks to people outside from activities within the laboratory or products and services generated by the laboratory. For example, the quality of analytical tests or synthesised products, control of the emissions of gases and liquids and the disposal of solid wastes.

To judge the performance against these standards it is necessary to have performance indicators, of which there are two types.

1. *Reactive Measures.* Monitor outcomes such as accidents and ill health. Examples include injury frequencies, lost time accidents and sickness absences.

2. *Active measures.* Monitor elements of the management process. Examples include numbers of people completing training in a specific topic or the percentage of measurements meeting a required standard.

The usefulness of a particular performance indicator can be judged by seeking answers to the following questions [B-23].

- Is it *relevant* to the business?
- Is it *amenable* to management intervention?
- Is it *simple* to understand?
- Is the data *easy* to collect?

3.2.2.3 Monitoring, Audit and Review

Monitoring of safety by management is an ongoing activity carried out frequently; say every few weeks, but at irregular intervals to ensure that staff do not do the right thing because the last Friday in the month is "inspection day". Any failings should pointed out to the staff concerned in a constructive manner at the time of the visit and not left to a later audit to pick this up. Spot checks during a casual visit are also very useful in ensuring that people working in the laboratory are behaving generally in a safe manner, e.g. wearing safety spectacles and other items of safety clothing. General safety and good housekeeping should be on the list of personal performance criteria for everybody working in the laboratory.

Having put in place adequate controls on the advice of a competent assessor, it is essential that R&D Managers arrange for an independent safety audit of their laboratories. These audits are designed to check that the written procedures are being followed and performance standards being met and not to question the validity of the standards. They are repeated much less frequently, say every two or three years, provided the function of the laboratory has not changed significantly during that period.

The review is a higher level managerial activity designed to see if the HSE programmes that have been put in place are achieving the desired objectives. If the answer is no then the review seeks to find out why this is the case. The objectives may be unattainable or just plain wrong in which case there is a need for a fundamental look at the policy behind these objectives.

This description, by its very nature, is only an overview on the main provisions of a Manager's role in ensuring a safe and healthy environment in the laboratory. There are many websites offering up to date information on HSE legislation [B-24], and innumerable training courses for those people working in companies who do not offer in house training.

3.3
Regulatory Affairs

Some regulations affecting R&D laboratories and chemical plants are statutory whilst others are voluntary.

The chemical, pharmaceutical and allied industries are highly regulated because of the hazards inherent in their operations, such as on their manufacturing plants and in their research laboratories. Pharmaceutical companies are subject to extra regulations because their products are intended for direct and widespread use by the general public, and hence it is extremely important that the products are safe, efficacious and of the highest quality. There are two types of regulations; those that are statutory and provide a "licence" to operate, e.g. Good Manufacturing Practice, especially important in the pharmaceutical industry, and those that are voluntary but need to be in place in order to satisfy a customer's demands for a quality service, e.g. ISO 9000 series.

In this section the regulations covering the operations within the chemical, pharmaceutical and allied industries will be covered in outline before going on to the concept of Total Quality Management. The other very important area of the regulations covering product registration will be described in Section D, 2.2.2.1.

3.3.1
Definitions of Terms

The worldwide standards adopted since the late 1980s is the International Organization for Standardisation (ISO) ISO 9000 series and the latest version (2000) contains definitions of relevant terms [B-25].

> *Quality:* The degree to which a set of inherent characteristics fulfils requirements.
> *Quality Management System:* The management system to direct and control an organisation with regard to quality.
> *Quality Assurance:* Part of quality management focused on providing confidence that quality requirements will be fulfilled
> *Quality Control:* Part of quality management focused on fulfilling quality requirements.
> *Product:*
> • A product may include service, hardware, processed materials, software or a combination thereof.

- A product can be tangible (e.g. assemblies or processed materials) or intangible (e.g. knowledge or concepts) or a combination thereof.

Quality Policy: The overall intentions and direction of an organisation related to quality as formally expressed by top management (ISO 9000: 2000).

Quality Management: Coordinated activities to direct and control an organisation with regard to quality (ISO 9000: 2000).

Quality Manual: A document describing the quality management system of an organisation (ISO 9000: 2000).

3.3.2
Quality Management Systems

One of the prerequisites for running a quality organisation is to have the necessary quality systems in place. These will be usually adopted at the company level and it is the job of the R&D Manager to implement, monitor and maintain the system within the laboratory environment.

3.3.2.1 The Quality Manager

All companies, whatever their size, need to appoint a Quality Manager, even if in smaller companies this is combined with other duties. Anyone with management experience may be suitable for the appointment, but in R&D the Quality Manger showed be suitably qualified, experienced and competent in the core technologies of the company.

The duties of the Quality Manager include the following:

- Issuing and controlling all the documentation in the QA system, including the Quality Manual.
- Training, or arranging the training, of all personnel selected to be internal auditors.
- Scheduling all the internal audits and ensuring the completion of all resulting corrective actions.
- Arranging management reviews of the quality system, drafting the agenda and ensuring minutes are taken, and that agreed actions are taken in a timely and effective manner.
- Advising all levels of management of changes to any quality standards to which the organisation is working and the responses they need to make.
- Maintaining contact with any bodies to which the organisation's quality system is registered.
- Representing the management on the occasion of any audit conducted on the organisation by a second or third party.
- Maintaining records of all quality-related activities.

3.3.2.2 Good Laboratory Practice (GLP)

Good Laboratory Practice (GLP) is a quality system concerned with the organisational process and the conditions under which non-clinical health and environmental safety studies are planned, performed, monitored, recorded, archived and reported. Adherence by test facilities to the principles of GLP ensures that there is a proper planning of studies and the availability of adequate means to carry out such studies. It facilitates the proper conduct of studies, promotes their full and accurate reporting, and a process whereby the validity and integrity of the studies can be verified. The application of GLP to regulatory studies assures the quality of the data generated and allows its use by Government regulatory authorities in hazard and risk assessment in particular of new substances. A very useful pocket-book guide to GLP and the UK GLP Regulations has been produced by the Department of Health, and should be consulted by students seeking further information [B-26].

The OECD has produced an internationally accepted set of "Principles of Good Laboratory Practice" in order to promote the development of quality test data. These have been ratified by the European Commission and transposed into EC Directives.

Comparable quality of test data forms the basis for the mutual acceptance of data among countries. If individual countries can confidently rely on test data developed in other countries, duplicative testing can be avoided, thereby saving time and resources. The application of these Principles helps to avoid the creation of technical barriers to trade, and further improve the protection of human health and the environment.

The Manager of a testing laboratory is responsible for ensuring that the Principles of Good Laboratory Practice are complied with in the facility. The OECD has set down the minimum responsibilities, which should be to

- Ensure that a statement exists which identifies the individual(s) within a test facility who fulfil the responsibilities of management as defined by these Principles of Good Laboratory Practice.
- Ensure that a sufficient number of qualified personnel, appropriate facilities, equipment, and materials are available for the timely and proper conduct of the study.
- Ensure the maintenance of a record of the qualifications, training, experience and job description for each professional and technical individual.
- Ensure that personnel clearly understand the functions they are to perform and, where necessary, provide training for these functions.
- Ensure that appropriate and technically valid Standard Operating Procedures are established and followed, and approve all original and revised Standard Operating Procedures.
- Ensure that there is a Quality Assurance Programme with designated personnel and assure that the quality assurance responsibility is being performed in accordance with these Principles of Good Laboratory Practice.
- Ensure that for each study an individual with the appropriate qualifications, training, and experience is designated by the management as the Study Director, before the study is initiated. Replacement of a Study Director should be done according to established procedures, and should be documented
- Ensure, in the event of a multi-site study, that, if needed, a Principal Investigator is designated, who is appropriately trained, qualified and experienced to supervise the delegated phase(s) of the study. Replacement of a Principal Investigator should be done according to established procedures, and should be documented.

- Ensure documented approval of the study plan by the Study Director.
- Ensure that the Study Director has made the approved study plan available to the Quality Assurance personnel.
- Ensure the maintenance of an historical file of all Standard Operating Procedures.
- Ensure that an individual is identified as responsible for the management of the archive(s).
- Ensure the maintenance of a master schedule.
- Ensure that test facility supplies meet requirements appropriate to their use in a study.
- Ensure for a multi-site study that clear lines of communication exist between the Study Director, Principal Investigator(s), the Quality Assurance Programme(s) and study personnel.
- Ensure that test and reference items are appropriately characterised.
- Establish procedures to ensure that computerised systems are suitable for their intended purpose, and are validated, operated and maintained in accordance with these Principles of Good Laboratory Practice.

3.3.2.3 Good Manufacturing Practice (GMP)

Good Manufacturing Practices, known internationally by the acronym GMP, are regulations that describe a set of principles and procedures covering the methods, equipment, facilities, and controls required for producing human and veterinary products, medical devices and processed food to the required quality. They are used by pharmaceutical, medical device, and food manufacturers as they produce and test products that people use. In the United States (U.S.), the Food and Drug Administration (FDA) has issued these regulations as the *minimum* requirements where they are called "current" Good Manufacturing regulations (cGMP), to emphasize that the expectations are dynamic [B-47]. Most countries have their own GMPs for drug and medical device manufacturers. GMPs define a quality system that manufacturers use as they build quality into their products. For example, approved drug products developed and produced according to GMP are safe, properly identified, of the correct strength, pure, and of high quality. A basic tenet of GMP is that quality cannot be tested into a batch of product but must be built into each batch of product during all stages of the manufacturing process.

At a generic level the GMPs adopted by most nations are very similar, with the set of basic requirements below.

- Equipment and facilities to be properly designed, maintained, and cleaned.
- Standard Operating Procedures (SOPs) be written, approved and followed.
- An independent Quality unit be established (like Quality Control and/or Quality Assurance).
- Both personnel and management should be well trained.

Managers need to be aware of the training requirements, which are quite specific. For instance in the EU the training requirements are as follows.

- The manufacturer should provide training for all the personnel whose duties take them into production areas or into control laboratories (including the technical, maintenance and cleaning personnel), and for other personnel whose activities could affect the quality of the product.

- Besides the basic training on the theory and practice of Good Manufacturing Practice, newly recruited personnel should receive training appropriate to the duties assigned to them. Continuing training should also be given, and its practical effectiveness should be periodically assessed. Training programs should be available, approved by either the head of Production or the head of Quality Control, as appropriate. Training records should be kept.
- Personnel working in areas where contamination is a hazard e.g. clean areas or areas where highly active, toxic, infectious or sensitising materials are handled should be given specific training.
- Visitors or untrained personnel should, preferably, not be taken into the production and quality control areas. If this is unavoidable, they should be given information in advance, particularly about personal hygiene and the prescribed protective clothing. They should be closely supervised.
- The concept of Quality Assurance and all the measures capable of improving its understanding and implementation should be fully discussed during the training sessions.

GMP contains ten principles that introduce employees to critical behaviours established by FDA and industry leaders to maintain good manufacturing practices in plants.

The ten GMP principles
1. Writing procedures
2. Following written procedures
3. Documenting for traceability
4. Designing facilities and equipment
5. Maintaining facilities and equipment
6. Validating work
7. Job competence
8. Cleanliness
9. Component control
10. Auditing for compliance

3.3.2.4 Quality Management Systems (ISO 9000 Series)
The worldwide standards adopted since the late 1980s are the International Organization for Standardisation (ISO) ISO 9000 series, the most recent version of which is ISO 9000: 2005. These are designed to ensure that customers are guaranteed the same quality when purchasing from a company in any country that has adopted these standards, provided that a company has the following:

- A quality policy
- Standardised processes of quality assurance
- A system of corrective actions
- Management review of the quality systems

Awarding ISO 9000 status to a company means that it has an effective quality management system. It does not mean that its products and services are of the quality demanded by the customer, and in particular does not address the technical requirements of a specified product.

Of the ISO series it is ISO 9001 that applies to the management of a laboratory. This requires that procedures be in place for all activities carried out within the workplace, that these procedures are recorded and that staff carries out the activities described in these procedures. Monitoring is required to ensure compliance and to modify and improve the procedures in the light of experience. The ISO 9001 Flowchart of the areas to be covered and supporting activities are shown in Figure B14.

Laboratory management is directly responsible for the following:

- *Work Plans.* Procedure by which the work plan and targets is constructed with the customer (sponsor) e.g. business unit, marketing or manufacturing functions.
- *Costs.* Procedure by which expenditure is agreed, monitored and reported back to the customer/sponsor.

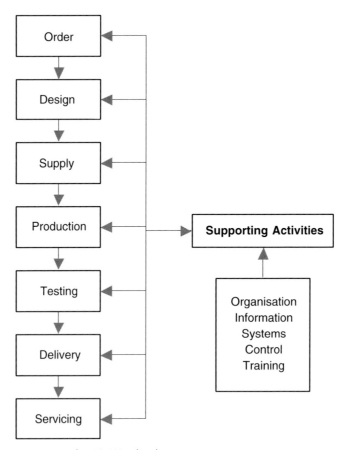

Figure B14. The ISO 9001 Flowchart

- *Reporting.* Procedure for recording work done for and reporting back to the customer.
- *Outcome.* Procedure by which the result of a project is agreed with the customer work plan, e.g. new product, process or termination.
- *Training.* Procedure by which training is identified and provided to staff in order to carry out the work for the customer to the highest standards.

The Quality System must be documented in a clear and efficient manner in the form required by ISO 9000, which must have the following three components:

> Part I. *Quality Assurance Policies*
> Policy Statement about what should and should not be done. Rules and principles governing the actions related to the policy. Underlines the objectives or guidelines for managing quality.
> Part II. *Procedures Manual*
> The manual provides precise details on how a particular procedure will be carried out to achieve the required quality. A procedure is a set of steps designed to meet a specified outcome.
> Part III. *Annexes*

All forms, reports, lists etc referred to during the quality assurance process.

Information systems must be in place to maintain the documentation and to ensure that it is up to date. In addition there is a need to maintain and store quality records and training records. This can be carried out within the Office function of the laboratory under the guidance of a Quality Manager.

Only a qualified company of independent experts can give recognition of the Quality System to ISO standards (Quality Audit). The system is subject to ongoing surveillance and assessment of its compliance with the standards. Regular inspection by trained internal auditors is also carried out to assess the company's continued ability to provide products and services of a suitable quality for the customer by asking the following questions:

- Are the procedures realistic?
- Are they appropriate?
- Are they followed correctly?
- Are the required results produced?

3.3.2.5 Environmental Management System (ISO 14 000 series)

The ISO 14 000 Environmental Management Systems (EMS) can be regarded as analogous to the ISO 9000 series, viewing the "customer" in this case as the environment, and just like ISO 9000, ISO 14 000 requires companies to "do what you say and say what you do." The relevant member of this family to chemical companies is the ISO 14 001 standard, which provides a framework on which to build a system that enables an organisation to meet the regulatory requirements. This is achieved by integrating environmental considerations into every level of the company and then being able to demonstrate the capability of the system through independent assessment.

R&D has an obvious and important role to play in any EMS, especially since competence in understanding the environmental impact of chemicals and processing is a high level requirement.

ISO 14 001 requires companies to develop and document a method for identifying, controlling, monitoring, and reviewing the ways in which the company potentially or actually interacts with the environment. To meet the standards of ISO 14 001 the EMS must address all of the following elements:

- *Environmental Policy.* This is a statement from top management that identifies the organizations commitment to continual improvement, prevention of pollution. There must also be a commitment to fully understand and comply with all relevant environmental legislation and regulations.
- *Planning.* Companies are required to identify ways in which they impact and interact with the environment and identify goals for improvement. In addition, companies need to identify all environmentally related legal requirements.
- *Implementation and Operation.* Documented procedures must be in place to control and monitor all of the ways in which the company impacts and interacts with the environment, including emergency preparedness and response. This includes an assurance that employees are trained and competent in their jobs that may interact with the environment.
- *Checking and Corrective Action.* Documented procedures that monitor and measure key characteristics of operations and activities that may impact the environment must be specified.
- *Management Review.* Requires top management to actively participate in reviewing the EMS.

Even though the focus of ISO 14 001 is on the establishment of an EMS and not specific performance levels, the outcome of having a strong EMS is improved environmental performance. Further information on ISO 9000 and 14 000 management systems can be obtained via the ISO website [B-27].

3.3.2.6 Data Recording

A key to following the requirements of GLP, GMP and other quality systems is having efficient, high quality recording of data. In the analytical sphere the advent of LIMS (see Section B, 2.9.1) provided the mechanism for logging approved data produced in the laboratory. In the synthesis laboratory the hand written notebook has held sway for many generations, largely because of legal and technical considerations. However, in recent years this has begun to change due to the ready availability of personal computers and the development of software for electronic notebooks [B-28].

Electronic notebooks have many advantages over paper-based notebooks, which include:

- They can be shared by groups of researchers, both locally and across geographical boundaries.
- Can be accessed remotely – accessible 24 hours a day and seven days a week.
- Cannot be misplaced, lost, or accidentally destroyed (if backed up).
- Legibility is better, e.g. does not suffer from unreadable handwriting.

- Easier to incorporate computer files, plots, graphs, etc.
- Coupling to laboratory instruments in remote locations is easy.
- Searching for information is easier.
- Can include multimedia – audio/video clips if desired.
- Can include hyperlinks to other information.
- Easier to distribute for communication purposes.
- Can be used to enforce legal and corporate standards.
- Report and scientific paper writing is made easier.
- Can be readily coupled into both LIMS and Document Manager systems for regulatory purposes.

Paper laboratory notebooks are extremely difficult to modify on the written page without leaving some evidence that the page has been modified illegally. Therefore, for centuries the legal system has recognised hand written records as the legal documentation of laboratory work and has accepted them as evidence of the date of and description of work performed in the laboratory. Electronic files on the other hand have been seen as are far too easy to modify and not acceptable to the courts, patent offices or the regulatory authorities. Very recently this attitude has changed. This is because there is now software that can notarise digital data in a way that provides indisputable proof that data has not been tampered with since it was notarised. These unique digital fingerprints are known as electronic signatures [B-29].

The two major questions to be answered by R&D management before there is the wholesale acceptance of the electronic management of data logging are as follows:

1. Is the system acceptable to the regulatory authorities?
2. Is the system acceptable to the patent offices?

In the USA the attitude of the FDA is paramount, especially to pharmaceutical companies. In 2002 the FDA, under its 21CFR part11 regulation, decided that electronic records are, under certain circumstances, equivalent to paper records and handwritten signatures. The driving force behind the use of electronic records is that they will be more cost effective for the pharmaceutical industry and the FDA to administer, having lower space requirements and being easier to retrieve than paper records. This should lead to a shortening of the time taken to get approval for a new drug. Although submission to the FDA using electronic records is currently voluntary, companies are trying to implement the rule as quickly as possible because there may be a time when the FDA will no longer accept paper records.

The new regulation has its biggest impact on analytical laboratories, which must meet the following requirements:

- Use of validated existing and new equipment and computer systems
- Secure retention of records to instantly reconstruct the analysis
- User independent computer generated time stamped audit trails
- System and data security, data integrity and confidentiality through limited authorised system access
- Use of secure electronic signatures for closed and open systems
- Use of digital signatures for open systems

Analytical and pharmaceutical laboratory management must respond to this rule by adapting their procedures to take account of the following:

- The current process of generating signatures has to be evaluated (who has to sign what and when).
- New procedures for authorised system access have to be developed in the company and in the laboratory.
- Existing computerised systems used for implementation must be updated with the correct functionality.
- The behaviour to use and handle I.D. codes and passwords as a basis for 'legally' binding signatures may have to be changed.
- New professionals such as an 'electronic archivist' may be required.

When it comes to the legal requirements for the protection of intellectual property the position is not so clear-cut. For instance in the US the PTO issues patents on the basis of the date of invention, i.e. the date in the laboratory notebook, not the date of application as is the case with other countries.

> "The legal community of internal legal counsel, third-party firms and the US Patent and Trademark Office (USPTO) is increasing its involvement in the issue. As the scientific community reacts to the availability of electronic laboratory notebook solutions, the pressure on internal counsel to form an opinion increases. Sometimes, the reaction of internal counsel is to seek external advice as well as look to the USPTO for possible rulings."
>
> "Several presenters at the Electronic Notebook Conference, held in January 2005 in Atlanta, discussed the viability of collecting laboratory data and information within a fully electronic environment, versus on the traditional paper medium. The consensus was that the case law has already been established. The basis of their arguments forces us not to look specifically for electronic notebooks being used in case law, but instead look for examples where electronic evidence has been submitted." [B-48]

However, it is claimed that the US PTO will now accept electronic notebooks provided the Federal Rules of Evidence are followed to the letter [B-29]. In spite of this many companies, at least for the near future, are continuing to maintain paper-based records, or a hybrid form, where electronic notebook pages are printed, signed and filed, for discovery research in order not to compromise any future patent applications.

3.4
Total Quality Management in R&D

TQM in no way conflicts with the long standing goals of an effective R&D Group.

Total Quality Management (TQM) has been part of our working environment for many years but has not diminished in its importance [B-30, B-31, B-32].

The following definition of TQM will be used as a basis for a discussion of its application to R&D.

In business terms, a Quality organisation is one that continually satisfies a customer's requirements, in both product and service. *Total Quality is* the goal of achieving Quality at the lowest possible cost. *Total Quality Management* is therefore the process of achieving *Total Quality,* by getting the commitment of everyone who works in the organisation [B-30].

The principles of TQM in no way conflict with the long-standing goals of an effective R&D Department. If R&D does not produce the right product or process, at the correct cost, in the shortest possible time, then it is not doing the job required by the company. Using the human resource, the scientific skills base, in an effective manner to achieve this quality product, should be the aim of every R&D Manager [B-31]. To achieve this quality the researcher needs to be asking the following questions.
- Am I asking the right questions to improve my fundamental understanding of a scientific problem often enough?
- Am I regularly solving the problem in the shortest time and at the lowest cost?
- Am I constantly providing the right new product or process opportunities for the customer?

In this very brief overview of TQM, as applied to R&D, three important aspects will be covered. The drawing up of procedures for the work activities, the process of continuous improvement and the methods used to compare the performance of the group with the best in the field, or *benchmarking* as it is known.

3.4.1
Quality Procedures

Most companies have set in place procedures to meet the requirements of the international quality standard e.g. ISO 9001 or national equivalents, as discussed in Section B, 3.2.2.3. The requirement of these standards are that procedures are in place for all activities carried out in the workplace, that these procedures are recorded and that the activities are carried out, by staff, as described in the procedures. The operation of these procedures is subject to an external audit, to confirm that there is compliance in the day-to-day activities of a work group. It should be noted that having procedures written down does not necessarily mean that an organisation has the best procedures in place. They will need to be constantly monitored and modified in the light of experience.

Typically the procedures for which an R&D Manager will be directly responsible, include:

- *R&D Targets.* The procedure by which these are agreed with the "customer", usually the marketing or manufacturing functions of the business. Evidence that the targets are actually being worked on and monitored.
- *R&D Costs.* The procedure by which R&D expenditure is agreed. The system by which costs are monitored and reported to the "customer".
- *Reporting.* The procedure by which all work is recorded and progress reported to the "customer".
- *Outcome.* The procedure by which the result of the R&D programme, for instance

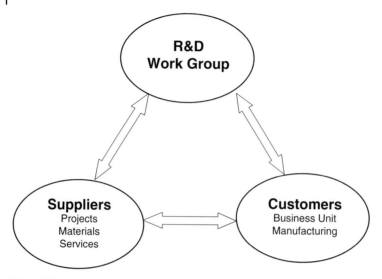

Figure B15. Customer–Supplier Relationship with R&D

a product or process, is agreed with the "customer". Conversely the method by which such a programme is terminated, in agreement with the "customer".

- *Training.* The procedure by which the necessary training for staff, enabling the work for the "customer" to be performed to the highest standard, is identified, provided and recorded.

3.4.2
Continuous Improvement

Having a set of procedures written down and in place is only the very first small step down the road to a Total Quality organisation. The key to going further along this road, is to recognise that all procedures and processes can be continuously improved. The activity of continuous improvement requires the involvement of all staff, at whatever level in the organisation. In this activity nobody is exempt, least of all the Manager, who must lead by example [B-32].

Three key groups of people who need to be involved in the process; the suppliers, the R&D work group and the customer. The suppliers include those people who provide the physical materials required for the job, and those who supply the services, for instance, information services to R&D. The internal customers for work done in R&D are the business unit, marketing or manufacturing, who are also its most important suppliers, supplying the targets for its work. Additionally, they will often supply a testing and evaluation service or a resource where this work can be done. Therefore, the process used in discussions with suppliers and customers can be considered as a cyclical one, as shown in Fig B15.

The observation that the customer is directly involved with both the inputs and outputs of the R&D work group is a key one. From this observation, a significant im-

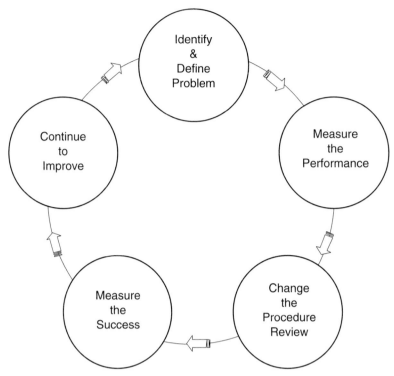

Figure B16. The Continuous Improvement Cycle

provement in the operation of R&D in the organisation can only be made with the direct involvement of the clients. The formation of joint problem solving teams is of paramount importance in this activity.

The process used for achieving continuous improvement involves the following steps:

1. *Identify and define the problem area.* Work closely with customers, usually for R&D, these are internal, e.g. marketing or production, but in the case of technical service they are more likely to be external, or "real customers" in the business use of the term.

2. *Measure the performance.* It should be obvious, to a person trained in the methodology of R&D, that there is a need to be able to measure and monitor activities, in order to be able to improve performance.

3. *Develop an improved procedure.* Select, either in consultation or jointly with the customer, an agreed new procedure from a list of alternatives drawn up by the Manager and the work group.

4. *Implement the procedure and re-measure the performance.* A figure, to be used as a measure of improvement in the performance, sufficient to justify the change in the procedure, should be agreed with the customer.

This methodology, known as the *continuous improvement cycle*, is shown schematically in Figure B16.

3.4.3
Performance Measurement Techniques

The methods and techniques, which are used to study and measure performance vary, depending on whether the problem is being identified or the causes determined, although there is a large degree of overlap. Most of these techniques will be familiar to people with a scientific background, especially those who work in R&D, for whom identification and analysis of problems is a daily activity. Some of the more commonly used techniques and where they are likely to be of use are listed below.

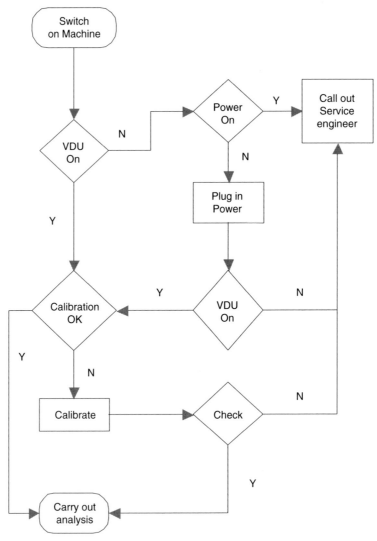

Figure B17. Flow Chart; Using an analytical HPLC machine

- *Flow Charts* are constructed to identify all the steps, or to define an ideal route in a process or activity. By this means deviations from the ideal can be identified and then worked upon. There are accepted symbols that are used to represent the individual processes in flow charts, which make it very easy to follow what is happening at any one point. An ellipse represents the start and finish, rectangles are process steps and a diamond is a decision point. To illustrate the method, a simple flow chart is shown in Figure B17 for the use of an analytical machine in the laboratory. In problem identification, a group of people with a high level knowledge of the process meet and carry out the following:
  1. Draw a flow chart of the actual process steps.
  2. Draw a chart of the ideal flow of the process.
  3. Compare the two charts, note differences for potential problems needing resolution.

- *Pareto Analysis* and charts. These are used to identify the major steps or defects in a process, the vital few rather than the trivial many. Based on the "80:20 rule" which states that, 80% of the problems are covered by 20% of the activities. As a metaphor "when clearing a rock fall across a road, move the large boulders first". It is easier to tackle the small problems but much less effective than dealing with the big ones. This is illustrated in Figure B18, where concentration on the area causing the largest numbers of defects, A, would be wrong because it is C which has the biggest effect on cost.

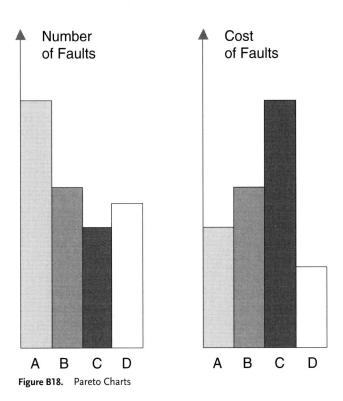

Figure B18. Pareto Charts

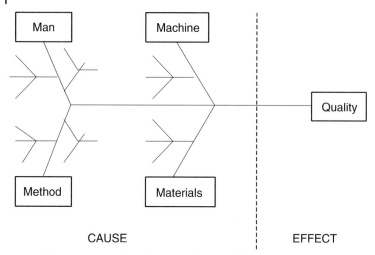

Figure B19. Cause and Effect Diagram (Ishikawa, Fishbone)

- *Cause and effect diagram.* This is also known as the Ishikawa diagram. This was introduced as a formal way of representing a specific *effect* and the possible *causes that* influence this effect. They are used to display all the possible causes of a specific problem and are usually constructed around two sets of four basic causes.
  1. Manpower, Machines, Methods and Materials (see Figure B19)
  2. Policies, Procedures, People and Plant

  Cause and effect diagrams, which are produced after group discussions on the problem, or its effect, involve the production of detailed check lists and a structured brainstorming (see Section C, 1.8.3). Following this process the diagram will become very complex and has the appearance of the skeleton of a fish; hence they are often called *fishbone diagrams.*

There are many other statistical techniques or graphical representations, commonly used in statistical process control on chemical plants and in laboratory experimentation, which can also be used in the context of continuous improvement. Typically these include *Run Charts* to display trends over time, *Histograms* to display the distribution of results, *Scatter Diagrams* to test for cause and effect by measuring one variable against another, *Cusum Analysis* to measure the deviation against an average or normal result against time and *Control Charts* to discover whether the variability is due to random causes (see also Section D, 3.4.2).

The use of these techniques will generate the data at the start of any improvement process. Once any changes have been implemented the techniques can then be used again to monitor the effects of these changes. Having confirmed the benefits of the change, the methods can be used to chart its progress over a period of time and also to indicate the next area for improvement. The process must be truly continuous if *Total Quality* is to be achieved.

3.4.4
Benchmarking

If good results seem to be emerging at regular intervals, it easy for R&D Managers to become complacent about the performance of their work groups. However, as a test of quality and to avoid any stagnation in performance, the question must be asked "how well are we really doing?" A method of checking the true or absolute performance of a group does not exist; it can only be done comparatively with another group. This involves comparing, using a range of parameters, the performance of the R&D group with that of another group, which is an accepted world class performer, in a process known as *benchmarking* [B-34, B-35].

Xerox first introduced benchmarking in 1979. Initially, it was used to compare the unit manufacturing costs and the features available in their copying machines with those from competitor companies. The process was expanded into a philosophy for ongoing improvement.

> "The continuous process of measuring our products, services, and business practices against competitors or those companies recognised as industry leaders." (Kearns, Xerox)

Benchmarking should not be seen as an alternative to TQM but as a parallel and complementary activity to TQM. There is little point in comparing the group's performance with another and then not doing anything about any deficiencies that are discovered.

There are four types of benchmarking:

1. Internal benchmarking. A comparison of internal operations.
2. Competitive benchmarking. A specific competitor to competitor comparisons for the product or function of interest.
3. Functional benchmarking. Comparisons to similar functions within the same broad industry or to industry leaders.
4. Generic benchmarking. Comparisons of business functions or processes that are the same regardless of industry.

The steps in the process of benchmarking are illustrated in Figure B20 [B-34]. In the planning phase, the first step is to choose what topic and which organisation is going to be used as the benchmark. It is easiest to start by making comparisons in the same industry sector and function whilst making some allowances for differences in size of the business and its organisational structure.

During the analysis phase the current performance gap is determined together with a projection into the future. At this stage, efforts are made to determine why an organisation is performing better than one's own. The McKinsey Group developed a model, which is designed for a comparative analysis of the innovation process, called the 7Ss model. The seven variables in this model are as follows:

- *Strategy.* Is there a product development strategy which defines types of projects selected and the resources required?

- *Shared values.* How much belief, enthusiasm and commitment Is there to innovation activity?
- *Style.* Is there top management commitment and how much support is there from the top for new product development?
- *Structure.* What lines of authority and responsibility are used for innovating activity?
- *Skills.* What specialist knowledge, tools and techniques are used for innovating activity?
- *Staff.* What is the involvement, empowerment, teamwork and degree of participation in decision making in relation to product development?
- *Systems.* What procedures, guidelines and control mechanisms are used for managing innovation activity?

Integration involves the communication of the results of the benchmarking analysis and then the setting of goals for the function. *Action* is what it says, the developing of action plans, their *implementation* and the *monitoring* of their impact. Maturity is when leadership is attained.

This comparison with outside companies is usually a top level or senior management activity. A newly appointed R&D Manager, whilst being involved, will probably only see a part of the total picture. At a practical level, it is useful for new or younger Managers to carry out an internal benchmarking exercise of their group. This will allow them to check on the performance of the group against the best elsewhere within the company's R&D. This can be done, for example, by seeking to find out how well the group is performing in comparison with Group Y, managed by a colleague, that is held in high regard by the general management within the company.

In this internal exercise, what meaningful measurements of performance can be used in order to make this comparison? Typically, the choice for obtaining metrics of an R&D group or laboratory can be selected from the following:

- *Innovation*
  - Patent applications filed
  - New Products/materials in the development pipeline
  - New application processes devised
  - Novel process chemistry introduced
  - New analytical procedures devised
- *Science*
  - Competencies
  - Papers published in top journals
  - Papers presented at Conferences
  - Awards obtained
- *Staff*
  - Numbers
  - Staff ratios – e.g. Managers-staff, PhDs-technicians
  - Working conditions
  - Costs
  - Safety – lost time accidents

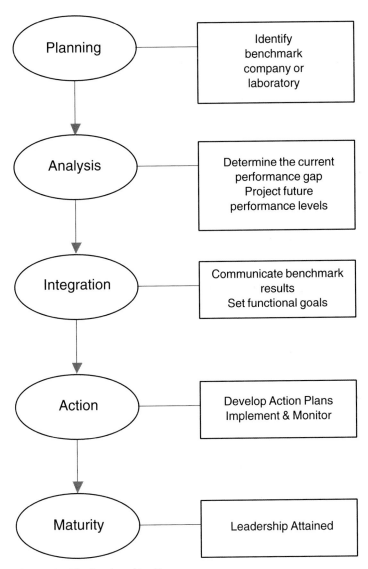

Figure B20. The Benchmarking Process

- *Infrastructure and Equipment*
  - Quality of building and services
  - Standard of equipment
  - Degree of automation
  - Flexibility

The data on all of these outputs should be measured on an ongoing annual basis. For a new Manager, taking over the management of an established group, where such data does not exist, it will be necessary to carry out a retrospective exercise. The data on the group chosen for the benchmark should not be obtained surreptitiously but in an open manner. After all, a compliment is being paid to the other Manager in considering that group as an example of high performance to be emulated by others.

The big disadvantage of the measures of performance in the list above is that, for the most part, they are only of internal relevance to R&D. The "customers" of R&D in the business function rarely show any interest in most of the professional elements on this list, for example the numbers of papers published or lectures given. To get continued and increasing support from the business team, there is a need to use those indicators that demonstrate the outcome of the group's efforts, in terms of financial benefits to the organisation.

It is often stated that R&D is there to provide the business with options for the future development. The final decision to launch a new product or to implement a new process involves considerations beyond the R&D dimension. Whilst this is no doubt the true position, it is only when some of those options lead to meaningful business for the company on a regular basis, that R&D is going to be held in high regard. In other words, there must be link between the expenditure on R&D and the business income.

The best information for an R&D Manager to collect is therefore that based on the eventual financial outcome of any completed work. There is inevitably a significant time gap between a product coming out of R&D and the actual generation of meaningful sales; this is covered in greater detail in Section D. Hence, the forecast sales income and profits of products in the development chain are often used to measure the outcome of R&D, both in terms of a new product and its attendant process. The Manager should keep an ongoing record so that actual growth in sales and profitability is recorded with time. There are some R&D activities in which the financial reward can be seen on a much shorter time frame. These include improvements to existing processes, the installation of new plant items and the introduction of new application processes for existing products. This type of hard data is invaluable; both as a measure of the performance of any R&D group, and in the longer-term justification of R&D spend by the company.

It should be remembered that benchmarking like TQM is a continuous process, as the competitors do not stand still. If R&D managers breathe a collective sigh of relief if they find that their group is pretty good in comparison with others, it will be short lived if the situation does not continue to improve.

3.5
Change Management

The process of changing from one state to the other state needs to be structured and well organized

Change management can be considered to be the task of managing change within an organisation. So-called change agents, who might be either internal or external consultants, can assist management in this task. Whilst underpinning the process of change is a body of knowledge, known as General Systems Theory (GST) [B-49].

R&D, by the very way in which it operates, has always been required to make changes in its working practices. If it does not do this it is highly likely that it will become stuck in its ways and lose its ability to operate at the highest technical level and fall behind competitor groups. Whilst the need for this type of change is obvious, e.g. the introduction of a new technology, there are many other instances where forces external to the research function, both from within the company and externally, such as new governmental legislation, require a more deep seated operational change within R&D. The effective management of such latter changes is especially important, but the principles behind change management can be applied in varying degrees on all occasions.

Some of the major changes that have had to be managed by R&D in recent years, many of which are ongoing, are listed below.
- Functions, to Matrix to, Strategic Business Units
- Business Process Reengineering
- De-layering and the impact on Succession Planning
- Mergers and Acquisitions
- Globalisation of R&D
- Move from 2nd to 3rd and 4th Generation R&D
- Quality Management
- ISO9000/14000, GLP, GMP
- HSE in the laboratory e.g. COSHH
- Performance Management, Performance Related Pay
- New Technology, LIMS, Automation, Electronic Notebooks, Informatics

3.5.1
The Change Process

If in the current state of the organisation there is position of "dissatisfaction" with an area of activity, i.e. it is unsatisfactory, then a move to a more desirable, future state is required.

Problems can arise when the change is broken down into the more manageable bits of the WHYs, WHATs and HOWs. When considering the HOWs it is easy to become hung up on one particular aspect. For instance asking, "how do we get the team to be more productive?" gets the thoughts centred on the means and not the goals of the process. WHAT questions centre on what we are trying to do, and have the danger of not questioning WHY we want to make this change. Thus the WHY question needs to be asked first and answered. "Why do we need to automate the lab?" "Because we need to improve our efficiency in line with our competitors". It is clear from this that the WHYs, WHATs and HOWs should to be integrated from the start of the process.

Therefore, when an organization needs to undergo any form of change some fundamental processes need to be adopted. Where you want to be needs to be analysed and understood. Getting the people on board who will be affected by the change is essential. The process of getting from one state to the other state needs to be structured and well organized, ensuring that any potential for disruptive impact during the process of change is avoided or kept to a minimum, but is always acknowledged by all stakeholders as a cost worth sustaining.

In all significant change initiatives there are three components that need to be in place:

A – There needs to be significant dissatisfaction with the CURRENT SITUATION – other wise not enough energy will be generated to alter that situation

B – There needs to be a clearly articulated view of the future intended situation – otherwise the rationale for the changes remains obscure. If that is the case then stakeholder understanding will be absent, and no one commits to something they don't understand.

C – There needs to be a way of getting from A to B

C is the Change programme and splits down into two cases:

C1 – where B is not that far from A, i.e. the envisaged situation is close to the current situation. In that circumstance C is a programme of continuous improvement – advancement by small discrete steps, with little or no disruption to ongoing operations.

C2 – where B is a long way apart from A, i.e. the envisaged situation differs really

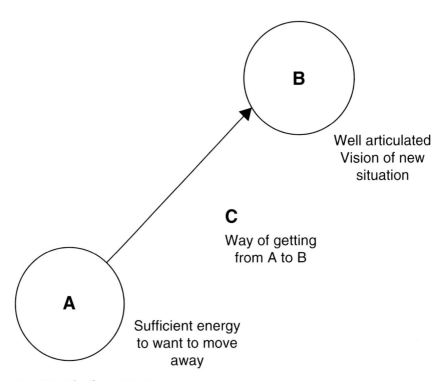

Figure B21. The Change Initiative

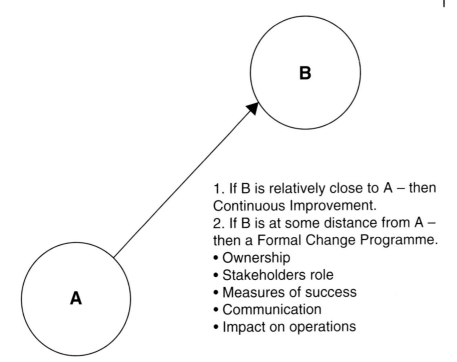

Figure B22. The Change Programme

significantly form A. Now C becomes a programme in its own right: properly struc-tured, owned, resourced, scheduled etc with its own plans and schedules, its own in-ternal measures of success and importantly stakeholder buy-in to the change activi-ties themselves. It is worth pointing out that in case C2 communications take on a crucial role. It is inadequate to issue a carefully thought out and precisely word-crafted letter describing B in positive terms. Stakeholders need to understand and commit to both B and C. That is they need to understand WHY we aim for B – and commit to working in B, but also they need to understand the change programme C and also commit to that. Some stakeholders will be happy with B but will not want to be in-volved in the changes needed to get there. Conversely some will agree to assist with C but will regard the ultimate goal B as not to their liking. It is important therefore to ascertain who of all the stakeholders fully understands B and C and to what extent they are willing to commit to either. Those who are 'happy' to be involved in the change programme but are quite unhappy to work in the way B will demand of them need special consideration e.g. alternative work is actively sought for them.

C2 may last a long time, say 12 months, and during that time some stakeholders will be operating both old and new ways of working in addition to being involved in the change programme C. It will be stressful and they will need to be supported. Also operational performance will unsurprisingly dip. This drop in performance needs to be negotiated with, and acknowledged by senior executives. C2 is also a situation where it is worth considering bringing in professional change expertise.

The choice of a strategy to enact any of these types of change will be governed in some way by the assets a manager has available to deploy on the project.

3.5.2
Change Management Skills

One leading authority on the process of introducing change in organisations has suggested that "Producing change is about 80 percent leadership – establishing direction, aligning, motivating, and inspiring people – and about 20 percent management – planning, budgeting, organizing, and problem solving".

He has also suggested that there are eight steps that a manager can take to transform their organisations [B-50].

1. Establish a Sense of Urgency
2. Form a Powerful Guiding Coalition
3. Create a Vision
4. Communicate the Vision
5. Empower Others to Act on the Vision
6. Plan for and Create Short-Term Wins
7. Consolidate Improvements and Produce Still More Change
8. Institutionalise New Approaches

Whilst these ideas were aimed at CEOs, they can be easily adapted to suit a more junior manager in R&D. Specifically managers will also have to demonstrate certain skills to make things happen within their own environment.
- Political
 There is often the need to strike bargains and negotiate through internal company politics when trying to institutionalise change.
- Analytical
 The ability to analyse the nature of the problem is essential
- People
 Change can only occur through people and the full range of interpersonal skills will need to be deployed
- System
 Such as those embodied in General Systems Theory
- Business
 An understanding of how the business works. This may be in relation to the R&D Group or the company, whichever is the most appropriate. Both are subject to having products, markets and customers as well as financial controls.

3.5.3
Embodying Change

Once a change has taken place in an organisation it is easy to heave a sigh of relief and assume that the manager's job is done. This is not the case until such change becomes embodied in the culture, as it easy for people to slip back into ways that they have been comfortable with in the past. The measurements of success, the critical success factors, need to be monitored and any deficiencies noted, and remedied as soon as possible; a process of continuous improvement.

As in Project Management (See Section D), a lot of new learning comes out during the process of carrying out the change. This body of knowledge needs to be captured and shared not only with members of the change management group but also with others in the organisation so that they can benefit from it when carrying out any changes in their areas.

References

[B-1] ROUSSEL, P. A., SAAD, K. N., ERICKSON, T. J., *Third Generation R&D: Managing the link to Corporate Strategy*, Harvard Business School Press, Boston, 1991

[B-2] GANGULY, A., *Business-driven Research & Development: Managing Knowledge to Create Wealth*, Macmillan, London, 1999

[B-3] MILLER, W., MORRIS, L., *Fourth Generation R&D: Managing knowledge, technology, and innovation*, Wiley, Chichester, 1999

[B-4] THAYER, A. M., Chasing the Innovation Wave, *Chemical & Engineering News*, Feb. 8, **1999**, 17–21

[B-5] PRAHALAD, C. K., HAMEL, G., Core Competencies and the Concept of the Corporation, *Harvard Business Review*, May-June, **1990**

[B-6] EIRMA Workshop Report, *Stimulating Creativity and Innovation*, Paris, 1993

[B-7] TROTT, P., *Innovation Management and New Product Development*, 2nd edn., Prentice Hall, Harlow, 2002

[B-8] BAMFIELD, P., Outsourcing- Now's the Time, *Chemistry in Britain*, 32 (17), **1996**, 23–24

[B-9] WHITE, R., JAMES, B., *The Outsourcing Manual*, Gower, Aldershot, 1996

[B-10] CAVALLA, D., *Modern Strategy for Preclinical Pharmaceutical R&D*, Wiley, Chichester, 1997

[B-11] www.ChemOffice.com

[B-12] McDOWALL, R. D. (ed.), *Laboratory Information Management Systems, Concepts, Integration, Implementation*, Sigma Press, Wilmslow, 1987

[B-13] TERRET, N. K., *Combinatorial Chemistry*, Oxford UP, Oxford, 1998

[B-14] NEWSOM, J. M. et al. in *Automated Synthetic Methods for Speciality Chemicals*, Hoyle, W. (ed.), Royal Society of Chemistry, Cambridge, 2000

[B-15] JENSEN, K. F. in *Automated Synthetic Methods for Speciality Chemicals*, Hoyle, W. (ed.), Royal Society of Chemistry, Cambridge, 2000

[B-16] http://www.cpc-net.com

[B-17] LEWIS, P. (ed.), *Health Protection from Chemicals in the Workplace*, Ellis Horwood, London, 1993

[B-18] *COSHH in the Laboratories*, 2nd edn., Royal Society of Chemistry, Cambridge, 1996

[B-19] PURCHASE, R. (ed.), *The Laboratory Environment*, Royal Society of Chemistry, Cambridge, 1994

[B-20] McCAIG, R., Hamington, M., *The Changing Nature of Occupational Health*, HSE Books, London, 1998

[B-21] HSE, *A Guide to Risk Assessment requirements*, HSE Books, London, 1996

[B-22] HSE, *Successful Health & Safety Management*, HSE Books, London, 1991

[B-23] WILSON, H. G. E., *Recent Developments in European Occupational Health Practice*, Proceedings of 14th Annual Conference of Australian IOH, Adelaide, 1995

[B-24] www.hsedirect.com and www.osha.gov

[B-25] BS EN ISO 9000: 2005 – *Quality Management Systems – Fundamentals and vocabulary*, www.bsi-global.com

[B-26] *The GLP Pocket-Book: The Good Labora-*

tory Practice Regulations 1999 and Guide to
UK GLP Regulations 1999, MCA Publications, London, 2000

[B-27] www.iso.org

[B-28] www.labtrack.com and www.cambridge
soft.com

[B-29] www.surety.com

[B-30] OAKLAND, J. S., *Total Quality Management*, Butterworth-Heinmann, Oxford, 1995

[B-31] TURNER, G. R., HADFIELD, R. P. (eds.),
.*Total Quality Management in the Chemical
Industry; Strategies for Success*, Royal Society
of Chemistry, London, 1994

[B-32] MCLAUGHLIN, G. C., *Total Quality in
Research and Development*, St. Lucie Press,
Boca Baton FL, 1995

[B-33] JURAN, J. M. (ed.), *Juran's Quality Control Handbook*, 4th edn., Mc Graw-Hill, New
York, 1989

[B-34] CAMP, R., *Benchmarking: The Search for
Industry Best Practices that Lead to Superior
Performance*, ASQC Quality Press, Milwaukee, 1989

[B-35] ROLSTDAS, A. (ed.), *Benchmarking: Theory and Practice*, Chapman & Hall, London,1995

[B-36] Dow, 2004, www.dow.com

[B-37] DuPont, 2005, www.dupont.com

[B-38] Pfizer, 2006, www.pfizer.com

[B-39] WYETH, 2006, www.wyeth.com

[B-40] Kalorama Information, *Outsourcing in
Drug Development: The Contract Research
Market from Preclinical to Phase III*, 2004

[B-41] www.innocentive.com

[B-42] MCCOY , M., C&EN, Jan. 26, 2004, 82 (4),
31

[B-43] EHRFELD, W., HESSEL, V., LÖWE, H.,
*Microreactors: New Technology for Modern
Chemistry*, Wiley-VCH, Weinheim, 2000

[B-44] EHRFELD, W., HARDT, S., LÖWE, H.,
Chemical Micro Process Engineering, Wiley-
VCH, Weinheim, 2004

[B-45] http://yet2.com/ip/microreactor.html

[B-46] C&EN, August 30, 2004, 82 (35), 9

[B-47] www.cgmp.com

[B-48] TORMEY, P., www.ip.com

[B-49] BENNIS, WARREN G., BENNE, KENNETH
D., and CHIN, ROBERT, eds. *The Planning of
Change* (2nd Edition). HOLT, RINEHART and
WINSTON, New York: 1969

[B-50] KOTTER, JOHN P. "Winning at Change"
Leader to Leader. 10 (Fall 1998): 27–33

[B-51] CHEESBOROUGH, H., *Open Innovation:
the new imperative for creating and profiting
from technology*, Harvard Business School
Press, 2003

[B-52] HARRIES-REES, K., Chemistry World,
February 2005, 34-39

Section C
Creativity and Innovation

It would be a very poor R&D organisation that was not both creative and innovative. That said, creativity and innovation are easily stifled or destroyed even in formerly effective groups. Like tender plants they only flourish in a suitable environment and are in need of constant nourishment, as is discussed in Section B. There are many actions that a Manager can take to help nurture the process of creativity and innovation and create the right climate in which chemists and other scientists can operate. To assist in this process the Manager needs to understand the characteristics of creative individuals and also the various pathways from creativity to innovative products or processes. A description of these and the techniques and aids involved forms the body of the first part of this Section.

The output from core R&D work is novel information, or intellectual property, as it is known. Having spent a great deal of time and money generating this intellectual property it is essential that it be protected against utilisation by a commercial competitor. This protection is commonly achieved by the filing of a patent. The timing of this activity, and the further development work required to support and strengthen the protection, is very important. Sometimes it is not possible, or appropriate to obtain patent protection for the results of R&D work. Companies can still exploit this sort of information, called "know-how", commercially. Prime examples of commercially valuable "know how" are the fine details in the manufacturing processes for chemicals. The protection of "know how" is much more difficult and secrecy is the only method. Additionally, such knowledge is often the basis for future work, it therefore needs to be properly recorded and safely filed.

The exploitation of intellectual property has to be considered early in the R&D cycle. The choice of the exploitation pathway materially affects the scale and type of R&D resource that will be needed. For instance, an opportunity may be further developed in house; by joint venture if one or more partner is needed; or by licensing the technology if it is outside the company's area of expertise.

The last few years have seen a much greater emphasis on the concept of the knowledge-based company. Knowledge is seen as the asset of a company and the R&D function as an important and essential component. The need to manage this knowledge for the benefit of the organisation has lead to the discipline of knowledge management and the requirements for this will be discussed in the final part of this Section.

Overview

Creativity represents the birth of ideas, innovation the process by which those ideas reach fruition. Everybody can contribute in some way to the creative process. Creative thinking and the generation of ideas need to be stimulated at both the individual and company level. Creative scientists are not identical; they come in all shapes and sizes. Innovation involves idea generation, translation to practice and commercialisation. Exploitation of the unexpected is a difficult activity but one which has the potential for high rewards. All staff can develop some kills in creativity by thinking laterally and by using specific techniques for idea generation. Free ranging, non-analytical techniques are the most successful. Select rather than reject the ideas generated. Intellectual property arising from R&D is an important part of a company's knowledge asset and needs protection. Develop IP in house, joint venture, license or file for the future. Managing knowledge efficiently and innovatively is a requirement in a knowledge-based economy. The tacit knowledge of one individual needs to become the explicit knowledge of all.

1
Creativity and the Nurturing of Innovation

This is a very complex area and includes contributions from most aspects of the business culture and management style [C-1]. Of the many factors, which can have an influence on the existence of a creative environment, and the nurturing of innovation, the most important is the climate or atmosphere within the company. Is it a company that fosters new ideas and is willing to act on them when they arise? Is it a company that encourages individuality, allowing creative people to express themselves with some freedom of thought? Does it encourage innovation to proceed in the various pathways to the market? If the answer is yes to these questions then it is in the hands of a Manager to deliver the rewards for the company.

1.1
Definitions

Creativity represents the birth of ideas, innovation the process by which those ideas reach fruition.

The words creativity and innovation are often confused, or used interchangeably, so it is worthwhile spending a little time clarifying the nature of and differences between these two activities.

In the Oxford English Dictionary, definitions of the verb *"to create"* include, "to bring into being" or to "form out of nothing". In a term more appropriate to R&D these could be redefined as *"making something new out of previously unconnected ideas or observations"*. Creativity is therefore the process by which a genuinely new idea is produced.

This definition of creativity shows that it is an active process. It may appear that creation happens by chance, but in the context of R&D it is usually a product of much prior knowledge and thinking. It is difficult to plan for the production of creative ideas. It is recognised that some people and organisations are more creative than others, but the reasons why this is the case are not easily defined. It is the job of R&D Managers to bring their group into this highly creative bracket. The chances of this happening can be greatly improved by a judicious study of the process by which creative ideas arise.

Research and Development in the Chemical and Pharmaceutical Industry, Third Edition. Peter Bamfield
Copyright © 2006 WILEY-VCH Verlag GmbH & Co. KGaA, Weinheim
ISBN: 3-527-31775-9

In the same dictionary, *"innovation"* is defined as "the introduction of novelties; the alteration of what is established" and "something newly introduced; a novel practice or method". We can re-interpret this for the industrial environment as *"the introduction of new products, application processes or services to the market based on prior knowledge"*. Thus, innovation in R&D is the generation of new ideas, which can then be used to produce these new products, processes and services. It should be obvious that this is an activity that does respond directly to planning. Certainly, at the development end of the innovation chain, rigorous planning is a prerequisite and this is covered in Section D on Project Management.

This black and white separation of creativity and innovation is fine in theory, but in practice is false. Creativity is required at all stages in the innovation chain leading to the development and introduction of new products processes and services. Creativity represents the birth of ideas, innovation the process by which those ideas reach fruition.

1.2
Total Creativity Management

Everybody can contribute to the creative process.

All functions within any successful business need to be innovative, but it is only R&D where this is paramount. R&D is the engine room for creativity and innovation within the company and this is its sole reason for existing. Consequently, if the right conditions for encouraging and exploiting creativity and innovation are not in place then the whole exercise will eventually be doomed to failure. Thus an understanding of the creative environment in which the competencies of the company are developed and exploited for its commercial well being is an essential for all managers having a technological role within an organisation.

For many years creativity and innovation was achieved in a rather ad hoc manner, only engaging those parts of the organisation as they were required, often in a linear way as this was seen as a logical progression. The more successful companies realised that this was inefficient. Blocks in the innovation pathway were only cleared as they appeared and often this lead to a resistance to change, particularly from those not involved from the start in the process of innovation.

Total Quality Management (TQM, see Section B, 3.4) is used by most companies to develop systems which can deal with the organisational problems related to the performance of the company, but this is inadequate when dealing with the generation of new and unusual ideas which need to be commercialised as rapidly as possible. A parallel activity is required on creativity. DuPont became the leading light in this area devising what is known as Total Creativity Management (TCM) [C-2]. This originated within the R&D functions but was so successful that they set up a Centre for Creativity and Innovation covering the whole company. The thinking behind this was that, unlike proprietary "hard" technology only available to the expert practitioner, creative thinking was a "soft" technology and hence could be shared between non-

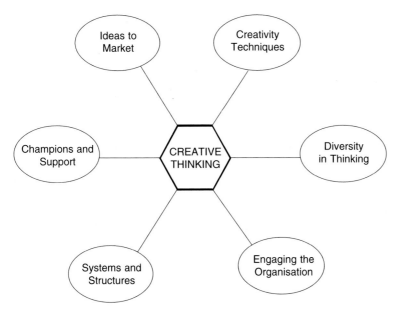

Figure C1. Six Dimensions of Creative Thinking

experts; everybody could contribute to the process, often in a challenging and unusual way.

TCM requires attention to six dimensions in the field of creative thinking. These
six dimensions, illustrated in Figure C1, involve the following:

1. Learning and applying creative thinking techniques that provide new ideas that
   seed creative thinking.
2. Capitalising on the value of diversity in thinking preferences and styles in building successful teams and task forces.
3. Engaging the organisation so that people will surface who have a clear idea of the
   values and of the initiative and want to take part.
4. Setting up structures and systems to sustain momentum in the initiative.
5. Recognising and rewarding emerging champions and supports.
6. Taking the best ideas to market, applying key components of the innovative process.

It is possible to draw up a matrix of the main factors that stimulate creativity and innovation [C-1]. Those factors affecting creativity centre on the individual and the environment whilst those influencing innovation are more concerned with teamwork
and organisation. There is a strong overlap in many areas, represented by the top
right hand box in Figure C2. The themes of this section will cover many of these factors, whilst others will be covered in the other sections.

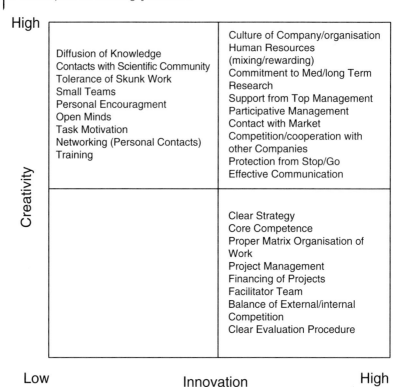

Figure C2. Factors Stimulating Creativity and Innovation

1.3
The Creative Climate

Creative thinking and the generation of ideas need to be stimulated at both the individual and company level.

Truly creative acts in science, as in life, are mostly unplanned but the provision of the "right" environment and climate are a great help in the process. The right environment does not necessarily involve a cosy and relaxed atmosphere. External forces posing a threat are often a stimulus to creative thinking, as has been experienced during the world wars of the twentieth century, when many of the great inventions influencing life in the latter half of the century were made or envisaged. It is also amazing how a group under threat of being disbanded suddenly start producing a raft of new ideas. The need for change to survive is expressed in the proverb, "necessity is the mother of invention". The arrival of a challenging technology that could threaten the survival of a whole industry is a great spur to creativity, as was experienced by the Swiss watch industry with the arrival of digital technology, and a similar effect was experienced by the photographic industry at the turn of the century.

Whilst the external climate may be hostile, the internal one within the R&D group must always be supportive for the creative individual.

Some of the most significant inventions, which seem to have happened by chance or serendipity, on closer examination often prove to have been the results of many years of endeavour and the accumulation of large amounts of data. In other words without the researcher having a goal and a chance to experiment, little would have happened. Why some people can work for years, accumulate vast amounts of knowledge, and still not be creative is another matter. Even the creative individual will under use this talent if the climate and stimulus for it to flourish are wrong for that particular person. Never underestimate the importance of the individual in the process of creation and innovation.

How can the Manager stimulate creativity? As in other aspects of management, there are factors that the individual manager can personally change or implement. The negative factors operating at a company level are very difficult to change or even influence, but recognising that they exist is a good start. Those factors external to the company must be taken as given, and Managers must work with these in mind at all times. The management style, to be used by R&D Managers with creative groups, has already been covered in Section A, 3.2. The factors that can stimulate creative thinking and the generation of ideas are as follows.

Stimulation at the individual level
- *Freedom to think.* Pressure must not be applied solely to achieve today's results. There should be sufficient time for the researcher to ask the questions "what if" or "why did that happen". Sitting down and thinking is a genuine research activity.
- *Chance to experiment.* The ability to carry out the unscheduled experiment is essential. Putting an idea to test is an important freedom in creative work. It is this type of experiment that often leads to serendipity, the genuinely unexpected result.
- *Wide contacts.* The putting together of unrelated facts to produce a new idea is aided greatly by discussions with other people. These can be other scientists, marketing colleagues and customers. Keeping creative scientists in an ivory tower is an intellectual imprisonment.

These three stimulating factors all require some "free time". The problem with most project management systems is that this is regarded as unproductive time and therefore something that can be sacrificed to meet other targets, for instance, be removed from the bar chart of a project plan. The Manager of a group, which is expected to be creative, will need to justify and fight for such "free time". This is not an easy task in an economic climate where companies are driven by cost/time considerations, but the arrival of the knowledge company, and the ideas behind it, is of great assistance.

Stimulation at the company level
- *Culture.* If the company's culture is one that has historically valued and rewarded the creative individual, this will clearly be beneficial.
- *Strategy.* The commitment to medium or longer-term research, leading to the exploitation of ideas as new businesses, products, processes and services, is integral to the company's R&D strategy. This implies management commitment and sup-

port at the highest level. It will instil confidence in the research establishment and gain support for its activities from elsewhere in the organisation.

These elements are seen as specific in helping to create the very special climate that allows creative work to flourish. However, there are several other factors which are common to both the creative and innovative climate, and these are described below.

1.4
The Climate for Innovation

Whilst, as already stated, it is almost impossible to separate creativity from innovation; most R&D work in chemical companies can be described as falling within the innovation process. The characteristics of an innovative climate are somewhat different and more specific than those for the creative one.

- *Clear strategy and targets.* The strategy is integrated into company policy; clear targets are set for R&D down to the individual business level, by a collaborative process. There is continuity of policy, the targets are based on sound business data and are not changed before R&D has time to attain the goal. This involvement of R&D in business policy and the target setting process is a big stimulant to innovation (see also Section D).
- *Correct skills base.* In most companies innovation is geared to improving or modifying an existing product range within an established business area. This means that there must be continuity in the skills base or the core competencies of the R&D community (see also Section A, 1). An intimate knowledge of what the competition is doing is very important if the innovations from the company are not going to be imitative and reactive. This knowledge can be obtained through a variety of means, including an ongoing study of the patent literature, up to date information on the introduction of new products and processes in to the market and knowledge gained at outside meetings.
- *Project management.* The formation of multi-disciplinary teams is very helpful to the innovation process. It brings together people with different strengths and knowledge, which they can then contribute to the process. Working towards a common goal against time constraints helps to produce the many innovative steps that are needed to get the correct product, process or service promptly to the market (see also Section D).
- *Strong interaction with the market.* This is a crucial requirement, especially in a rapidly changing market environment. For example where a new technology is being introduced. Many leads are obtained, by discussion and observation on the customers' premises, which will stimulate innovation to the benefit of the customer and hence the company. These visits by R&D scientists to the customer should not be rare events. They should be integral to the process of innovation.

1.5
The Creative and Innovative Individual

Creative scientists come in all shapes and sizes.

All R&D Managers want to have their groups staffed by people who have ideas. Many will agree with the old maxim that R&D needs such people "even if their ideas do not work"; at least they are producing ideas, and the big one may be just around the corner. Thus at the core of any creative or innovative step lies the contribution made by individuals. These creative ideas people are often not the type of persons who can turn their ideas into innovations or can function well in time driven project teams. It is interesting that Japan, for so long the powerhouse of economic developments, reacted to its economic difficulties in the mid 1990s by questioning its educational system. The Japanese educational system has produced a generation of graduates, who have been excellent in the era of the mass produced and the uniform high quality production systems that have been pioneered in that country. This uniformity is good at innovation but performs less well in creativity terms and in the ability to respond to rapid change. Getting the correct balance between conformity and individuality is seen as a major challenge to the country's educational system.

The personalities of creative scientists, or the inventor types of person, mean that they are much more likely to change their behaviour in response to new ideas and processes. This means that they are much happier working in new areas or topics rather than within old or existing ones, where they can easily become bored. They like to change tack in the light of new data and so will be frustrated working a well mined seem. In these areas, their ideas are likely to be met with comments from the established practitioners such as, "we tried that five years ago and it did not work".

There is a widely held view that the creative individual is introverted, only interested in the internal world of ideas and has poor social skills. The experience of the author, over many years in an industrial research establishment, is that the opposite was more likely to be the case. Inventors tended to more extroverted in their behaviour than the average person, sometimes bordering on the eccentric. Social skills may have been be somewhat lacking but interpersonal relationships with colleagues were rarely a problem. It is the "academic" chemist who is more likely to be the introverted character. Interested in ideas for their sake rather than what they can produce for the organisation. The recruitment of introverts is a bad policy for an R&D Manager to follow. As far as creative scientists are concerned it seems to be the case that, you know one when you work with one, but nothing is more certain than that they will all be different.

A list of dimensions, or attributes, can be drawn up for the innovative scientist. These are the people who will form the greater part of the professional staff within R&D. Some of these attributes have already been discussed in Section A, 1.2 for the recruitment process.

- *Results oriented, with a strong innovative drive.* This person wants to achieve a successful outcome to the work and likes to work on the leading edge of science or technologies.

- *Concerned about standards and efficiency.* Wants the work to be always done to the right standard. Likes to find ways of doing things better than in the past and to improve on the techniques involved.
- *Takes risks and shows initiative.* These risks are usually calculated ones and not gambles. They are ones that are designed to take the process on to the next step.
- *Seeks information and is inquisitive.* Looks for the information to solve the problem and to generate new ideas. Likes to look at the process personally, an observer seeking first hand knowledge. Wants to know why things happen.
- *Conceptual and analytical thinker.* When problems arise seeks an answer by putting forward a new paradigm for consideration. Uses analytical thinking to justify new ways of looking at and overcoming problems.
- *Forward looking.* Thinks for the future, stays with a problem until an answer appears. Stores ideas for future use.
- *Strong at influencing the key people by rational argument.* Knows the key people who need to be influenced and acts to gain their commitment, using data and logical arguments which are soundly based.
- *Concerned with impact of ideas on others, communicates well.* Shows concern about the impact any ideas may have on own professional standing.
- *Is independent, with a good self-image and self-confident.* Is not afraid to raise problems or to stand up for own ideas. Aware of own scientific and technical ability as well as the limitations. Confident enough to challenge existing assumptions.

Being capable of finding problems, generating ideas and *solving problems* are among the key skills of creative individuals; being intuitive and analytical as well as practical. Other characteristics include: observant, inquisitive, lateral thinking, can suspend judgement, can handle and clarify complexity [C-1].

Because Invention involves venturing into the unknown, individuals with nonconformist characteristics and who are keen to work outside traditional parameters in order to achieve a better result are likely to be most successful. Such people are more likely to *take calculated risks* in order to model the future and be least affected by social pressures to minimise risk and avoid change.

Creativity and innovation carry with them the need for the individual to be able to *communicate* their new ideas to others. Presenting their ideas fluently, both verbally and/or in writing, to other members of project teams during the innovation process is essential if they are to be acted upon. Conversely the individual must be able to listen to other people, understanding and responding to their feelings, needs and concerns about new ideas and changes of procedures.

Creative individuals often dream up certain solutions, products of *fantasy* that question conventional wisdom. However, dreaming must not be the sole activity and should not impede the ability to make specific decisions when required. *Optimism,* is also a notable characteristic of the creative scientist, believing that "there might be a solution", where the pessimist would rather say, "there might not be a solution".

There are many factors that are likely to inhibit the creative or innovative individual. Probably the most important of these has already been discussed elsewhere and this is the fear of failure. Failure for the individual can take many forms; it can be seen as losing credibility in the eyes of the peer group or even worse, fear of being laughed at by colleagues for apparent naiveté; it can be a fear of failing in a chosen career, indicated by a lack of promotion or a poor financial reward. These fears will always exist, but they will only be really meaningful in a climate that punishes failure, of whatever form. An R&D Manager must make sure that this climate does not exist within the group.

Creative individuals tend to dislike the boring repetitive tasks far more than the average person. This is especially true if these tasks are of a non-scientific or bureaucratic nature. A thoughtful R&D Manager will try to protect the more sensitive members of staff from these relatively mundane activities. It is important to check that this attitude is not just a way of avoiding the less satisfying tasks within the group, in the hope that they will fall on to the shoulders of other members. If this is allowed to happen it will lead to a build up of resentment within the group.

1.5.1
The MBTI Creativity Index (MBTI-CI)

Having a metric or personality test that could define accurately a creative individual would be of great help in placing people in the correct job. There is one tool, developed by Harrison Gough, which is designed to perform just such a function [C-3]. It is called the MBTI Creativity index (MBTI-CI) and is based on the Myers-Briggs Type Indicator (MBTI). MBTI, the most widely used instrument for understanding normal personality differences, was developed by Isabel Briggs Myers and her mother, Katharine Cook Briggs over 50 years ago [C-4]. The MBTI instrument determines preferences on four dichotomies:

- *Extraversion (E)–Introversion (I)*. Describes where people prefer to focus their attention and get their energy; from the outer world of people and activity or their inner world of ideas and experiences.
- *Sensing (S)–Intuition (N)*. Describes how people prefer to take in information; focused on what is real and actual or on patterns and meanings in data.
- *Thinking (T)–Feeling (F)*. Describes how people prefer to make decisions; based on logical analysis or guided by concern for their impact on others.
- *Judging (J)–Perceiving* (P). Describes how people prefer to deal with the outer world; in a planned orderly way, or in a flexible spontaneous way.

Combinations of these preferences result in 16 distinct personality types. Understanding characteristics unique to each personality type provides an insight on how they influence an individual's way of communicating and interacting with others.

Taking a person's MBTI score and placing them into a formula, developed over many years of research, calculates the MBTI-CI. From this research Gough was able to suggest that creative individuals tend to be more intuitive (N) rather than sensory (S), more perceiving (P) than judging (J), more extroverted (E) than introverted (I) and more thinking (T) than feeling (F).

The average creativity index figure is 235.5, in a potential range of 84.5 to 547.5. Gough concludes that people with scores less than 250 are less likely, whilst those above 350 are more likely to demonstrate breakthrough creativity.

1.5.2
Kirton Adaption-Innovation Inventory

Another widely used metric is the Kirton Adaption-Innovation Inventory. This is based on the Kirton Adaption-Innovation Theory, which considers that, before people can feel comfortable when tackling a problem they will differ amongst them on what structure they require and whether there is a need for a consensus on that structure. Another important factor is how important is the outcome of the problem solving, what is the reward or alternatively the punishment to be felt by the people concerned. This is called the "cognitive style" [C-50].

The key assumption of this theory is that there is a continuum of cognitive style, ranging from high adaption to high innovation, on which people can be placed dependent on the characteristic mode in which they solve problems (create or make decisions). The Kirton Adaption-Innovation Inventory (KAI) is the psychometric measurement devised to locate respondents on this continuum. It is concerned with the different ways in which people think, and particularly the way they show their creativity, solve problems and make decisions. Three sub traits, which are measured to give a total score, are explained as follows [C-51]:

- 'SO' (Sufficiency of Originality): Those more adaptive prefer to produce a sufficiency of ideas within an existing framework which immediately seem plausible; the more innovative tend to be more obviously radical in their style and to produce a proliferation of ideas whether or not they are needed.
- 'E' (Efficiency): the more adaptive have a preference for thoroughness and attention to detail; the more innovative typically have broader ranging views, incorporating elements often thought irrelevant, who tend to start many things and then get bored following them through. Their style helps them break paradigms, being less wedded to the detail of any one structure.
- 'R' (Conformity to the Rules and the norms of a Group): The more adaptive prefer working within existing practice and custom, valuing group cohesion. They tend to solve problems by good use of rule; the more innovative may act as catalysts to the group and often seem to solve problems by bending or even breaking rules.

Some defining characteristics of adaptors and innovators found using this psychometric measure are shown below [C-51].

Adaptors. Safe, reliable, methodical. Disciplined and efficient. Masters detail. Prefers defined problems. Rarely challenges the rules; solves problems by use of rule. Seeks consensus, values group cohesion. Does things better. Provides balance when working with innovator.

Innovators. Thinks in risky, unexpected ways. Little respect for past custom – seen as irrelevant. Trades off detail for over-view. Questions definition of problem. Often challenges the rules; solves problems despite rule. Can appear insensitive, even abrasive, to group cohesion. Does things differently. Provides dynamics for radical change.

1.6
Innovation Pathways

Innovation involves idea generation, translation to practice and commercialisation. Before proceeding we need to re-address the question "what is innovation?" There have been many attempts to answer this apparently simple question. Drucker says that it is the act that endows resources, including existing resources, with a new capacity to create wealth; changes the value and satisfaction obtained from resources by the customer; the successful exploitation of change [C-5]. From a more practical viewpoint, Akio Morita (Sony) considers that neither science alone nor technology alone is innovation, because science provides us with information, which, though previously unknown, only offers hints at the future, and technology comes from employing and manipulating science into concepts, processes and devices. True innovation is made up of the following three inclusive and necessary factors combined with a nurturing environment.

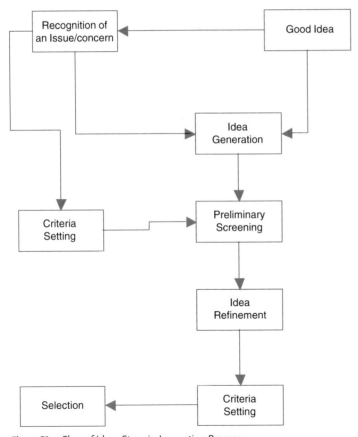

Figure C3.   Flow of Ideas Stage in Innovation Process

- Creativity in technology
- Creativity in product planning
- Creativity in marketing

Innovation is extremely important to any company, as it is the key to growth. Internally it can invigorate the whole structure of the company and is a strong motivator to employees. Externally, without innovation a company's product line stagnates, its customers become dissatisfied and eventually the business dies.

The innovation process involves three basic steps.

1. Invention or idea generation
2. Translation
3. Commercialisation

The idea generation step is key. The idea can come from the recognition of a concern or issue to the company or a customer or can come from somebody having what appears to be a bright idea. All ideas need to be evaluated by a screening process in an iterative process, as illustrated in Figure C3 [C-6]

The translation step involves the development of products and processes, whilst commercialisation involves the implementation of an innovation in a form acceptable to the identified market and then its diffusion to other areas. These aspects are covered in much greater detail in Section D.

All R&D Managers should recognise that innovation within commercial companies is not restricted to the R&D function. Indeed in the most effective companies, innovation is integral to the activities of staff in all the functional components.

1.6.1
Sources of Innovation

Exploitation of the unexpected is a difficult activity but one which has the potential for high rewards.

There are many ways in which innovations can arise and be brought to the market within companies and Drucker exemplifies seven such sources [C-5].

1. *Areas of Unexpected Innovative Opportunity*
   - Unexpected Success, e.g. the discovery of the nylons (DuPont).
   - Reaction to unexpected failure, e.g. temporary adhesives (3M).
   - Reaction to unexpected outside/adjacent event, e.g. development of ink jet printing for use with the home/office computer lead to the need for new high performance dyes and pigments.
2. *Incongruities that provide Innovative Opportunity*
   - Between the economic realities of an industry, e.g. move from bulk scale plants to speciality chemical plants.
   - Between reality and assumptions, e.g. shipping industry went for faster more economic ships but should have targeted time in port as key economic criterion, thus container ships and Ro-Ros.

- Between the analysis of an industry and the values and expectations of its customers, e.g. IBM and personal computers.

3. *Process Need as an Innovative Opportunity*

   For example one area well known in the chemical industry is process de-bottlenecking. Another example is the need for more environmentally friendly processes. Drucker's example is plate glass making by the float glass process (Pilkingtons).

4. *Indicators of Impending Change in Industry Structure*

   Drucker lists four scenarios where this might happen:

   1. When there is rapid growth.
   2. When there is a doubling in the size of an activity.
   3. The convergence of hitherto separate technologies.
   4. If the way its business is transacted is changing rapidly.

   Examples of this are the late nineties phenomenon of divestments, mergers and of take-overs in the oil, chemical and pharmaceutical industries; the focus on core competencies and the extensive use of subcontracting/outsourcing.

5. *Opportunities from Demographics*

   Drucker's example uses changes in population (i.e. size/young adults the third world); age structure especially more elderly people; distribution between town and country; more women in employment; more in further/higher education; higher disposable income. With little analysis consequences are predictable. For chemistry opportunity lies with medicines for the aged e.g. anti Alzheimer's disease, life style drugs and nutriceuticals.

6. *Opportunities from Perception*

   Drucker's examples are health, dining and food, care for environment, class structure and feminism. There are many chemical examples here particularly under care for the environment e.g. CFCs replacements, unleaded petrol, City diesel, solar electricity and fuel cells.

7. *Knowledge based Innovation*

   Knowledge-based innovation is characterised by the longest lead-times and they are almost never based on a single factor, e.g. a new plastic requires input from organic synthesis, catalysis, structural characterisation and chemical engineering. They have to fulfil a unique set of requirements to be successful i.e. a careful analysis of all necessary factors: a clear focus on strategic position (complete system or market or key function, because you seldom get a second chance); the need to learn and practice entrepreneurial management. One of the best examples in the last few decades was the invention of biphenyl liquid crystal materials by George Gray and his colleagues in the UK in the early 1970s. Although these were rather crude by today's standard they never the less lead to the worldwide development of liquid crystal displays, a multi billion dollar industry.

Complementary to Drucker's Approach is one based on the analysis carried out by Magrath of inventions brought to the market by companies. From this analysis six basic pathways were identified [C-7].

1. Innovation based on core technologies
2. Innovation based on a unique mix of common operating elements

3. Innovation that satisfies unmet customers needs
4. Innovation created from pure imagination
5. Innovation based on scientific research
6. Innovation based on functional excellence

Those innovations of type 2 are more usually found in the retailing and service businesses; type 4 in creative organisations in the arts, publishing and communication industries rather than in mainstream chemical companies. The other four types are all found in the Chemical and Pharmaceutical Industry, and more specifically in the R&D functional groups.

Innovations based on core technologies involve the company taking its existing core technology and expanding its application into new areas. The company is building its business on the base of its core competencies. The often-quoted example of a company building on its core technology is Honda. The technology or competency at the core of their business was small engines. Honda used this competency to introduce a wide range of innovative products into the market, including automobiles, motorcycles, lawn mowers, and chain saws.

> A simple example, which illustrates this approach in the chemical industry, was the one adopted a few years ago by the nuclear fuel company, BNFL (British Nuclear Fuels) in the UK in the early 1990s. This company had as one of its core technologies from within nuclear power context, the ability to handle fluorine on the bulk scale. It used this expertise to diversify into the manufacture of fluorochemicals. One use for these organic chemicals was as intermediates in the production of biologically active chemicals and related materials. This fine chemicals business was a very different one to that of nuclear fuel.

In attempting to diversify, companies have to get involved in building new competencies, for instance in the above case the production and marketing of fluorochemicals. This is sometimes a long and arduous process. Recruiting appropriately trained chemists and acquiring a marketing function already working in the area is often used to shorten the time.

For these new ventures to be successful, and to find new avenues for the exploitation of its core technology, the company needs to make an ongoing commitment to R&D. R&D will be required to build new competencies to meet the challenge of the diversified business. The implication for R&D management is that it must have a forward plan to match their resource and technology with the company's strategy.

Innovation which satisfies unmet customer needs requires R&D to have a deep involvement with customer companies in existing and related market areas. It is best exemplified by the finding of new uses for existing products and thus expanding the market. The R&D work is very closely related to the technical service activity with customers, and is often carried out by the same people, especially in the speciality chemicals sector. Because the R&D often involves the development of new or improved applications for existing materials it is sometimes known as *"applications research"*.

An example of this type of innovation is given in the Magrath paper [C-7]. It involves Church & Dwight, a manufacturer of Arm & Hammer baking soda in the US. This company for many years expanded its business in sodium bicarbonate by satisfying customer needs in many areas. These included the deodorising of refrigerators, cleaning of drains, polishing silver, reduction of toxic emissions from gas power plants, neutralising acid in swimming pools, and in toothpaste.

Sometimes ideas for new applications are generated by the general public writing to companies with a request for a product to carry out a particular task. An R&D Manager who is responsible for application research within a company should have access to any letters that arrive. Even though 99.9 in 100 of these ideas will be of nil value, or even from cranks, sometime or other one of them could provide the germ of an idea.

Innovation based on scientific research needs little justification as a way of producing innovations to research chemists. It is the reason R&D departments exist and is the main source of innovation within the chemical and pharmaceutical industry. The conversion of the results of the work on research targets into products for the market, or of laboratory experiments into manufacturing processes, are the paths that most R&D workers will follow. There are well-tried methodologies that can be used to ensure that the milestones along the way are met, in a timely fashion. These will be described in Section D on Project Management.

The most difficult innovation for a company to handle is one that arises because of an unexpected result or chance observation, serendipity as it is often called. The fact that it is unexpected means that the system is not prepared to exploit the finding. It needs a person with a strong entrepreneurial streak to exploit this type of finding; otherwise it might be lost amongst the more conventional results. The birth of the synthetic dyestuffs industry in the middle of the 19th century, and arguably that of the organic chemical industry, was based on a serendipitous event, and provides a classical example.

> W. H. Perkin in 1856 was attempting to make quinine, whose structure was unknown, by the oxidation of allyl toluidine with potassium dichromate in sulphuric acid. This reaction failed so he tried the oxidation of aniline, again no quinine was produced but he observed that extraction of the crude product with ethanol gave a purple solution, which deposited crystals on cooling. Perkin considered that this could be used as a dyestuff and indeed this turned out to be the case. Commercial dyers were approached and it was found to dye silk in a superior way to the then used natural dyes. It was later named Mauveine. Perkin had been particularly fortunate in that the formation of Mauveine requires ortho and para toluidine, as well as aniline, and these were present as impurities in his sample of aniline. Whilst this was an immensely important discovery it was the exploitation of this observation by Perkin that led to the innovation of synthetic dyestuffs. Over subsequent years he was able to scale up the processes for aniline and Mauveine and to sell it commercially. A method was found to apply the new dyestuff onto wool, greatly expanding its sales, this latter finding being an innovation of the type which satisfied unmet customer needs. This story has many parallels throughout the history of chemistry and other sciences [C-8, C-9].

The message to any researcher and Manager is: exploitation of the unexpected is a difficult activity but one which has high rewards when successfully achieved. The results provide a genuinely novel opportunity for the company to exploit the sole rights of the associated intellectual property.

For the Manager *innovation via functional excellence* is not very different from that achieved using R&D's core technologies or competencies to the benefit of the company. These core skills can also be used to contribute to a novel method of using functional excellence in another part of the company. For instance, R&D could have a particular skill that could be used as part of an innovative service package to customers. It is common to use the functional excellence of R&D in the development of comprehensive environmental packages that are built up by companies as part of their service to customers. The carrying out of environmental audits and the provision of advice on the efficient use of energy in customer's plants which utilise a company's chemicals are examples of more recent innovations.

1.6.2
Business Analysis Techniques

Before staring to innovate it is obviously advisable to have a full understanding of the competitive business environment in which the company is currently operating. What sectors is it in, what are the strengths in the key areas, where are our weaknesses, who are the competitors and what are they doing? The results of this analysis can be used not only to define where competencies are lacking but also areas where there is an urgent need for innovations. They are therefore useful as a base line for idea generation.

Two of the analytical methodologies that are commonly used to assist in this process have the acronyms STEEP and SWOT. STEEP, which stands for Sociological, Technological, Economic, Environmental and Political, is an extended version of PEST, both of which involve the analysis of these key drivers in a specific location or on a global scale depending on the type of organisation concerned. A brief example of some of the components of a typical PEST analysis is shown in Table C1.

SWOT is the business analysis technique used most commonly by companies and other organisations. SWOT stand for Strengths, Weaknesses, Opportunities and Threats. It is usually set out in quadrants on a single sheet of paper as illustrated in Figure C4. The contents of the SWOT analysis are usually generated by a brainstorming technique (see Section C, 1.8.3).

Table C1 The PEST Analysis

Political	*Economic*	*Social*	*Technological*
Tax policy	Economic growth	Health	R&D activity
Employment laws	Interest Rates	consciousness	Automation
Environmental	Exchange rates	Population	Technology
regulations	Inflation rates	Growth	Incentives
Trade restrictions		Age distribution	Rate of technological
Tariffs		Career attitudes	change
Political Stability		Safety consciousness	

Strengths   Weaknesses

Patents Strong brand names Good reputation with customers Cost advantage from proprietory know-how Exclusive high grade natural resources Favourable distribution networks	Lack of patent protection A weak brand name Poor reputation with customers High cost structure Lack of access to natural resources Lack of distribution channels
Shifts in customer taste from company products Emergence of substitute products New regulations Increased trade barriers	An unfilled customer need Arrival of new technologies Loosening of regulations Removal of trade barriers

Threats   Opportunities

Figure C4.
SWOT Analysis

Whilst the SWOT technique is most commonly used for business analysis, where potential competitive advantage can be identified from the fit between the company's strengths and upcoming opportunities, it can also be used in creative idea generation. Looking at the Opportunities and Threats in a more divergent rather than an analytical mindset is likely to produce some ideas, which can be exploited in an innovative manner.

1.7
Creative Thinking

Everybody can develop some kills in creativity by thinking laterally and by using specific techniques for idea generation.

Creative thinking is generally viewed as an activity carried out by an individual, in silent contemplation. Whilst this is the way most creative thinking is done it can also be carried out by a group of people. In his seminal work, "Lateral Thinking for Management", Edward de Bono challenged the idea that creativity was a magic gift, that nothing could be done about it unless you possessed this magic gift [C-10]. He stated that everybody could develop some kills in creativity by thinking laterally, not in logical straight lines. In other words creative thinking can be improved by using specific techniques for idea generation. It is in the context of group activity that creativity tools are most useful. In a group, ideas expressed by one individual can generate new trains of thought in others, and putting some structure into the process can assist this interplay.

Creativity arises from what appear to be illogical or even irrational thought processes. Such thought processes are easily and invariably inhibited by negative external factors. An individual can choose the time, place and the topic for creative thinking, whereas in a group this is not possible, the wishes of the individual must fit in with the rest of the group. Barriers to creative thinking in the individual can occur in such environments. Three types of barrier, which can apply to a group of people as well as an individual, can be identified.

1. *Perceptual.* The person sees only part of the truth, does not analyse all the factors involved.
2. *Social.* A desire to conform to the prevailing environment.
3. *Emotional.* The fear of being ridiculed, the loss of personal status, the wish to come to a rapid conclusion.

An R&D Manager must be take these points into account when using any of the techniques described later. Steps must always be taken to minimise the impact of perceptual, social and emotional barriers on the activity.

The process of creative thinking is obviously very complex but can be usefully simplified into the following steps.

- *Preparation.* This is the stage when data as facts or experience is accumulated. Analysis and a mental building of models occur.
- *Incubation.* This is work at the subconscious level and therefore nothing is physically recorded.
- *Insight.* The sudden flash of inspiration. This occurs during periods of positive thinking, never when thinking negatively or when under critical examination. Such insights need to be recorded promptly or they will disappear as quickly as they appeared.
- *Verification.* This is where ideas are judged and evaluated against known facts and proved. A lengthy and laborious process.

In the *preparation* stage, many of the analytical or problem identification techniques, described in the paragraphs on the continuous improvement aspects of TQM, are very useful. These methods include Ishikawa Fishbone Diagrams, Pareto Charts, and Flow Charts etc. (see Section B, 3.4).

It is common practice within R&D to use tools to aid or stimulate the insight stage [C-11]. When using such techniques the principal objective must be to create the conditions, climate and environment, which are conducive to people having insights into a problem. Therefore, to ensure success when using any of the methods it is essential to separate the *evaluation procedure* from the process of *idea generation.*

1.8
Tools for Creativity

The desire for companies to be creative in all aspects of their businesses has been paralleled by the arrival of many management texts with creativity in their titles. Just as many others have been written by consultants specialising in teaching or applying creativity tools to a whole range of industries. There are web sites devoted to creativity that are useful resources, having details of books, tools, courses and links to other relevant sites [C-12]. It should be noted that most of the tools in the following sections are now available in software versions. These are known collectively as Creativity Software [C-47].

It is impossible to insist on creativity but it is possible to assist in the process by employing the right tools. R&D Managers should be familiar with such tools, understand when they can be used most effectively and encourage their use within teams, groups and individuals.

1.8.1
The Generation of Ideas

Free ranging, non analytical techniques for the generation of new ideas are the most successful.

In R&D there are some problems which need to be answered within fixed parameters and therefore restrictions have to be placed not only on the number ideas which can worked upon but also how far those ideas can stray from known pathways. However, there are many other occasions when there are no or very few restrictions placed on the ideas which can be followed in order to resolve a problem.

The generation of new ideas is key to opening up the vista for innovative research, so any methodology which helps the process is welcome and valuable. Over the last forty years an extensive and growing list of techniques for creative thinking have been devised, as illustrated in Table C2 [C-12].

The discussion in this section will concentrate on the free ranging, non analytical techniques for the generation of new ideas that seem to have found the widest and successful application, namely lateral thinking, brainstorming, six thinking hats, morphological analysis, synectics, mind mapping, rule reversal and TRIZ as well as considering the distinctive, alternative approaches of Robert Fritz.

1.8.2
Lateral Thinking

For most of our education we are taught to think in a logical manner. This is called vertical thinking because the process involves proceeding directly from one state of information to another, and in this sense is continuous. Lateral thinking is a discontinuous process and depends on the ability to escape form the rules of vertical thinking. The essential differences between the two types of thinking are given in Table C3 [C-10].

Lateral thinking is something that is used sparingly. In fact the vast majority of the time we use vertical thinking. Lateral thinking comes into its own when we want to break out of the mould, to take radical new directions. Both types of thinking are used in the process of creation. Lateral thinking turns up an idea, vertical thinking develops it.

One of the most potent aspects of lateral thinking is the use of provocations. These are statements that challenge our perceptions in a provocative way lifting us out of traditional, literal ways of seeing a problem. These include "reversals" and "distortions" An example of the use of reversals is provided by work at DuPont [C-2]

> A fibre plant technical group was dealing with the issue of: "How to improve continuity of our complex continuous flow filter system?" The filter system, an important step in polymer manufacture, was based on a reciprocating belt with 70 moving parts that had frequent failures. This caused costly disruptions in the plant and problems in product quality. The group manager was a creativity champion whose group was educated in lateral thinking. The "reversal" technique led to this provocation: "The moving belt is stationary." This provocation shifted thinking to an entirely new direction and led to design of a system that reduced the number of moving parts by 80%. The result was a major breakthrough in process continuity, improved product quality and significant cost savings.

Table C2   Some Major Creative Thinking Techniques

Technique	Source
Lateral Thinking	Edward de Bono, *Lateral Thinking for Management* [C-10]
Random Input	E. de Bono, *Teach Your Child How to Think* [C-37]
Six Thinking Hats	E. de Bono, *Six Thinking Hats* [C-15]
Problem Reversal	Charles Thompson, *What a Great Idea* [C-38]
Ask Questions	– ditto –
Brainstorming	Alex Osborn, *Applied Imagination* [C-39]
Forced Relationships	Robert Olsen, *The Art of Creative Thinking* [C-40]
Attribute Listing	M. Morgan, *Creating Workforce Innovation* [C-41]
Morphological Analysis	Fritz Zwicky, *Discovering Invention etc* [C-42]
Mind Mapping	Tony Buzan, *Mindmap Book* [C-18]
Synectics	William Gordon, *Synectics* [C-17]
Lotus Blossom	Michael Michalko, *Thinkpak* [C-14]
Idea Toons	M. Michalko, *Thinkertoys* [C-13]
Realm of the Senses	Mike Vance, *Think Out of the Box* [C-43]
Use of Drawing	Robert McKim, *Experiences in Visual Thinking* [C-46]
NLP	http://nlp.com/NLP
DO IT	Robert Olsen, *The Art of Creative Thinking* [C-40]
LARC Method	R. Williams & J. Stockmyer, *Unleashing the Right Side of the Brain* [C-44]
Basadur Simplex	www.helicon.com/simplex
TRIZ	Genrich Altshuller, *Creativity as an Exact Science* [C-20]
Fuzzy Thinking	Bart Kosko, *Fuzzy Thinking* [C-45]

1.8.3
Brainstorming

Brainstorming, devised originally by Alex Osborn, is the most regularly used and valued of all the methods. It is a technique that is used to generate a large number of ideas on a selected topic, and is carried out informally and lacking any real discipline, in the complete absence of any critical process or evaluation of the ideas generated. Brainstorming is therefore not suitable for use on a topic where one correct answer, or one that is immediately applicable, is being sought. It is used when a break out from a current area is required or when a start is being made in a new one of a non specialised nature.

It is a group activity, optimally between five to ten people. At least two-thirds of the people invited to the brainstorming session should have ideally a variety of experience obtained in different fields. Preferably an experienced, neutral facilitator should run the session, whose job is to record the ideas on a flip chart, electronic white board or other appropriate technology and to stimulate and orchestrate the flow of ideas from the participants. For somebody inexperienced in facilitating a brainstorming session there are some guidelines to be followed.

Before the session
- *Advance warning*. Tell people what the subject is some days before the session, providing briefing notes if appropriate, to allow ideas to incubate.
- *Environment*. Choose a place where people will feel relaxed, preferably away from the workplace. Do not use oversized rooms and arrange the seating so that people can see each other.
- *Timing*. Allow sufficient time for the discussion but have a clear end point. Use prime time in the mid week, never as a Friday afternoon filler.

During the session
- *Rules*. Ensure that you and the participants are aware of and follow the rules of brainstorming.
- *Suspend judgement*. Do not allow criticisms to be raised during the session.
- *Freewheel*. Do not put the brakes on any discussion.
- *Quantity*. Get as much material as possible out of the meeting, encourage participation from everybody.

Table C3   Vertical versus Lateral Thinking

Vertical Thinking	*Lateral Thinking*
Chooses	Changes
Uses YES/NO	Does not use YES/NO
Uses information for its meaning, is analytical	Uses information to set off ideas, is provocative
One thing follows from another	Makes deliberate jumps
Concentrates on what is relevant	Welcomes intrusions
Moves in the most likely directions	Explores the least likely directions
Is a closed procedure	Is open ended

It is the job of the facilitator to keep firm but unobtrusive control of the session and to make it enjoyable. If the discussion flags ideas must be injected.

Osborn identified nine principle ways of manipulating a subject, which were rearranged by Bob Eberle into the mnemonic SCAMMPERR.

Substitute something
Combine it with something else
Adapt something to it
Magnify or add to it
Modify it
Put it to some other use
Eliminate something
Rearrange it
Reverse it

One good way of following these principles is to use deliberately challenging or stimulating checklists. The list used commonly consists of the following "forcing" verbs (Osborn).

Magnify	Reverse
Minimise	Combine
Rearrange	Rotate
Alter	Multiply
Adapt	Divide
Modify	Transpose
Substitute	Unify
Add	Subtract
Distort	Complement

Another checklist asks challenging questions (Van Fange).

- What about shape, size?
- What if reversed, inside out, upside down?
- What else can it do?
- What can be left out?
- What if carried to extremes?
- What if symmetrical? Asymmetrical?
- Can it be safer?
- Slide instead of rotate?
- Can it move? Can it be stationary?

These check lists are used by the facilitator to stimulate questioning of a particular aspect. This helps to promote associations, encourage adaptation, modification or the substitution of ideas by the participants.

There are also available games and other aids, which have been devised to stimulate thinking during a brainstorming session and other idea generating methods. Examples include the toys and card games devised by Michael Michalko and a range of stimulation techniques gathered together by Brian Clegg and Paul Birch [C-13, C-14, C-15].

It is important that the facilitator stifles any criticism of the ideas being generated, whenever it occurs, no matter the context. Disruptive flippancy by any of the participants is also a negative influence on the proceedings and is to be avoided. Most of all the facilitator must be on the look out what are called "killer phrases". These are phrases, which, although common in normal conversations, can in this context inhibit the flow of ideas.

"We tried that years ago."
"Not a bad idea, but it won't work in this place."
"Scientifically unsound."
"Far too expensive."

Or those generated by self-doubt in the participants.

"This may be a stupid idea, it may not work."
"Perhaps I have not thought this idea through."
"You may not like this suggestion."

During the session it is absolutely imperative that every idea is recorded for subsequent analysis, as they are made and without debate. The simple method of writing the ideas on a flip chart is the most effective; it is quick, cheap and does not convey a permanency or ownership, which is advantageous in the freewheeling atmosphere of the session. However, electronic white boards are being used Increasingly. Another alternative is to have laptop computers networked together in which the ideas are typed in anonymously during the discussion. Whatever method is employed, do make sure they are used fully as it is no good asking people to recall their ideas in the cold light of morning, when they can be dismissed or more likely disowned.

1.8.4
Synectics and Metaphors

Synectics is from the Greek word *synectikos* defined as the *"bringing things of a different nature into connection"*. Originally this was taught by Synectics Incorporated as a procedure for creative thinking through the use of analogies. In this method analogies are used to install in the mind concepts which normally would not be connected by a logical thought process [C-16].

There are three fundamental precepts of synectics theory [C-17].

1. Creative output increases when people become aware of the psychological processes that control behaviour.
2. The emotional component of creative behaviour is more important than the intellectual component; the irrational is more important than the intellectual component.
3. The emotional and irrational components must be understood and used as precision tools in order to increase creative output.

A method closely related to synectics is one called *Metaphorical analogy*. As its name implies it uses one problem and its solution as a metaphor for a problem in a completely different area, which has not yet got a solution. It is simpler to use than synectics and is possibly better when dealing with a specific problem.

Returning to synectics, the procedure is followed in two parts.

1. Making the strange familiar.
2. Making the familiar strange.

The *making of the strange familiar* is done in preparation for the creative thinking phase. In this stage the problem is broken down into its components, relevant data are collected and a model is constructed to show the inter-relationship of the unsatisfactory parts. This is designed to identify the "real" problem. It is then possible to restate the problem. At this point, known or likely solutions are written down but left for evaluation against the new ideas.

The *making of the familiar strange* activity is then performed. This involves looking for analogies to the problem and considering how things are handled in these cases. Three types of analogy are sought.

1. *Personal.* That means self-identification with a component of the system. What is your role in the system? It is a bit like method acting, "what does it feel like to be a tree?"
2. *Direct.* The team is asked to think of a completely different situation where the same phenomena or symptoms exist. The task is then to see if they can be applied to the current problem. This is much the easiest area for scientists and technologists to find analogies. There are many examples in the history of scientific innovations. The classic, often quoted example, is that of Brunel's invention of the caisson for underwater construction which was based on his observation of shipworms tunnelling into timber.
3. *Symbolic.* From the redefinition of the problem, carried out in the making the familiar strange phase, it is possible to look at attributes or keywords in the description to see if class-characteristics can be identified. The task is then to see if these characteristics suggest any new answers to the problem.

This is not an easy method to facilitate without prior training. The facilitator of the sessions should not take any direct part in the development of analogies. The Manager's task is to stimulate the team by asking provocative questions in a similar way to brainstorming. The objective is to keep the team on the right track and prepare them to move on to another one, if that being followed seems to be leading to a dead end.

1.8.5
Mind Maps

Making associations between one image or word is an important activity of the brain when a person is thinking. Each image or word has innumerable links to other words and images leading from one idea to another. Mind maps, developed by Tony Buzan, are designed to model the way the human brain works, i.e. by associative and linear processes, and so assist in the generation of new ideas [C-18].

A mind map is made by starting in the centre of a page with the main idea or problem, and then working outwards in as many directions as possible creating new nodes

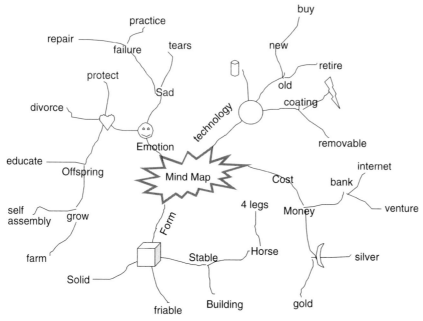

Figure C5. Mind Map

at each association point, composed of keywords or images. The structure then grows in an organic fashion producing a mass of associations, as illustrated in Figure C5.

It is important to work quickly and to use key words, and not phrases and draw the images in colour to highlight the new nodes. Using this methodology the mind forms associations almost instantaneously. Mind maps have great creative potential when used as part of a brainstorming session.

1.8.6
Morphological Analysis

Morphological analysis is a good tool to use as an aid in the setting of long term R&D targets either in new business areas, new processes or designs. It is not very useful in answering a problem in short-term innovation targets. Using this method it is possible to generate a large number of ideas in a very short time.

The initial step in the method involves breaking the problem down into the important functional areas, dimensions or sequential process. This method is best illustrated using an example. For the purpose of this exercise let us consider the area of liquid crystal displays in the electronics industry (Table C4).

Even this very simplified version of the LC Display area produces $3 \times 5 \times 3 \times 5 = 225$ combinations. By applying additional business constraints, for instance time to market, cost, profits etc. it is easy to see how the number of combinations can rapidly increase.

Table C4   LC Displays (Morphological Analysis)

LC Materials	LC Properties	Circuitry	Type
Monomeric	Twisted nematic	Simple matrix	Direct
Polymeric side chain	Super twisted nematic	X-Y Matrix	Projection
Polymeric backbone	Chiral smectic	Active Matrix	Flat panel
	Ferroelectric		Large Area
	Anti-ferroelectric		High definition
	Tuneable birefringent		
3	5	3	5

The next step in the procedure is to look at all these combinations and list them against the following criteria.

1. Known
2. Would work but needs further study
3. Completely impossible in present state

The items coming from point 2 provide definite areas for work within R&D and justify further evaluation. It is also possible that those from point 3 might justify deeper thought and consideration for long-term research.

1.8.7
Six Thinking Hats

The Six Thinking Hats, designed by Dr. Edward de Bono, is an ingenious framework to think through a subject in a focused way that makes time and space for creative thinking [C-19]. It is used extensively in companies such as DuPont, IBM, Prudential Life Insurance, British Airways and Siemens to have efficient, productive meetings.

The underlying principle in the Six Thinking Hats framework is that parallel thinking is more productive than argument. There are six symbolic hats as shown in Figure C6. Each hat has a different colour and represents a different dimension in thinking about the subject being addressed. Everyone wears the same hat at the same time. The hats can be put on and taken off depending on the sequence of thinking that makes the most sense. The White Hat deals with information, the Red Hat with feelings, the Black Hat with caution, the Yellow Hat with benefits, the Green Hat with creative ideas, and the Blue Hat with thinking about the thinking process.

The Six Thinking Hats can be used either individually or in groups. The worked examples below illustrate its use. The first (A) takes a hypothetical human resource issue, the second (B) concerns introducing a new business [C-2].

(A) The Company wishes to introduce a four-day week for all staff. It gathers representatives from within work groups and across them to discuss the issue using the Six Thinking Hats procedure.

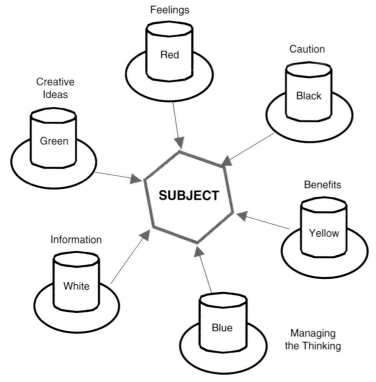

Figure C6. Six Thinking Hats

1. *Blue Hat*
   Facilitator managing the process, thinking about thinking.
   Choreographs the procedure, decides which hats to be worn.
2. *White Hat*
   Information, data and needs.
   Responses: Which four days? How long each day? All staff the same four days?
3. *Red Hat*
   Feelings, intuition, emotion.
   Responses: Might not see my lunch companion. Will need to change my car club.
4. *Yellow Hat*
   Logical, positive, benefits.
   Responses: Will come refreshed. More time for hobbies. More time with family.
5. *Black Hat*
   Judgement, caution.
   Responses: Dealing with customers difficult. Productivity drops.
6. *Green Hat*
   Creative, new ideas.
   Response: Could have two days on, two days off in a cycle.

(B) A strategic planning team in an R&D Division supporting a portfolio of business units developed a concept that could lead to a profitable new business. The new business would broaden the portfolio, capitalising on strengths of existing businesses. The concept was controversial. The business managers felt the plan would dilute resources from their businesses and strongly opposed the proposal. The R&D management was strongly in favour. Upper management scheduled a two-hour meeting to resolve the issue. The R&D team organised the meeting based on the Six Thinking Hats framework, which was familiar to the group.

> The meeting started with a brief White Hat overview of the concept and discussion to clarify the proposal. The facilitator then had everyone put on the symbolic Black Hat. This created an energetic discussion among the business managers resulting in many flip charts listing serious difficulties inherent in the concept. This approach allowed the business managers to air all the reasons, many justified, why they were opposed to expanding the portfolio with this new business.
>
> The facilitator then had everyone switch to Yellow Hat thinking. There was dead silence. The R&D people purposely said nothing. Then one of the business managers noted a benefit. Then another business manager noted another benefit. Within 15 minutes all of the business managers joined in and there were as many flip charts on benefits as there were on difficulties.
>
> Next, the facilitator moved to the Green Hat to generate ideas on how to retain the benefits while overcoming the difficulties. By this time everyone was energetically generating ideas to upgrade the proposed concept to make it workable.

The outcome was that the R&D team was charged to develop an implementation plan for further review embodying the changes suggested by the business managers who now strongly supported the proposal.

1.8.8
TRIZ

TRIZ is a problem solving method that was developed by Genrich Altshuller over a forty-year period in the former Soviet Union [C-20], it has many devotees, used in many major companies and there is an on-line journal devoted to the topic [C-21]. The acronym is derived from the Russian for Theory of Inventive Problem Solving. It is very different from the other creativity techniques described above, in that it operates via a study of patterns of problems and solutions and not by the spontaneous creativity of individuals and groups. It was based on the analysis of over 1.5 million patents, since extended to 2.8 million, to discover patterns that predict breakthrough solutions to problems.

What Altshuller did was to remove the subject matter from a patent to uncover the problem solving process. There were three primary findings from this research of the patent literature.

1. Problems and solutions were repeated across industries and sciences
2. Patterns of technical evolution were repeated across industries and sciences
3. Innovations used scientific effects outside the field where they were developed

From these studies he was able to categorise the solutions to problems into five levels.

- *Level one.* Routine design problems solved by methods well known within the specialty. No invention needed. About 32% of the solutions fell into this level.
- *Level two.* Minor improvements to an existing system, by methods known within the industry. Usually with some compromise. About 45% of the solutions fell into this level.
- *Level three.* Fundamental improvement to an existing system, by methods known outside the industry. Contradictions resolved. About 18% of the solutions fell into this category.
- *Level four.* A new generation that uses a new principle to perform the primary functions of the system. Solution found more in science than in technology. About 4% of the solutions fell into this category.
- *Level five.* A rare scientific discovery or pioneering invention of essentially a new system. About 1% of the solutions fell into this category.

He also noted that with each succeeding level, the source of the solution required broader knowledge and more solutions to consider before an ideal one could be found. This is consistent with the need for more creative thinking as we go up the levels.

What Altshuller postulated was that over 90% of the problems engineers faced had been solved somewhere before. If engineers could follow a path to an ideal solution, starting with the lowest level, their personal knowledge and experience, and working their way to higher levels, most of the solutions could be derived from knowledge already present in the company, industry, or in another industry.

Altshuller found that each of the most inventive patents primarily solved an 'inventive' problem, which he defined as those which contain conflicting requirements, which he called 'contradictions'.

An example of a technical contradiction is provided by an automobile airbag, which needs to deploy very fast in order to protect the occupant (good) but the faster it deploys the more likely it is to kill small people (bad). An example of a physical contradiction is provided by coffee that should be hot to be enjoyable but cold enough not to scald the customer.

He defined 39 basic properties and 40 principles for solving problems containing contradiction in any two-of–39 properties (see Tables C5 and C6).

By extracting and organising the most frequently occurring contradictions and the principles of the resolution of these contradictions, a matrix of 39-improving parameters and 39-worsening parameters (39 × 39 matrix), with each cell entry giving the most often used (up to four) of the 40 inventive principles, was created. This matrix is known as the *contradiction matrix*. This matrix is the simplest and the most straightforward of all the TRIZ tools, and also the one most commonly used. A small corner of such a matrix is shown in Figure C7.

Altshuller, whilst compiling the data for the contradiction matrix, also found that the evolution of various technical systems was not random but in fact followed objective laws. He found that the evolution of any system could fit into one of eight specific patterns of technical system evolution, which are given below:

Table C5   The 39 Features of TRIZ

Number	Feature
1	Weight of moving object
2	Weight of stationary object
3	Length of moving object
4	Length of stationary object
5	Area of moving object
6	Area of stationary object
7	Volume of moving object
8	Volume of stationary object
9	Speed
10	Force
11	Stress or pressure
12	Shape
13	Stability of the object's composition
14	Strength
15	Duration of action by a moving object
16	Duration of action by a stationary object
17	Temperature
18	Brightness
19	Use of energy by moving object
20	Use of energy by stationary object
21	Power
22	Loss of Energy
23	Loss of substance
24	Loss of Information
25	Loss of Time
26	Amount of substance
27	Reliability
28	Measurement accuracy
29	Manufacturing precision
30	External harmful affects the object
31	Harmful side effects
32	Ease of manufacture
33	Ease of operation
34	Ease of repair
35	Adaptability or versatility
36	Device complexity
37	Complexity of control
38	Extent of automation
39	Productivity

1. Life cycle of birth, growth, maturity and death, e.g. steam engine and propellers replaced boats with oars.
2. Trend of increasing ideality, e.g. printers with better resolution and printing speeds.
3. Uneven development of sub-systems resulting in contradictions, e.g. powerful aero-engines developed faster than the wing design.

Table C6   The 40 Inventive principles

Number	Inventive Principle
1	Segmentation
2	Extraction
3	Local quality
4	Asymmetry
5	Combining
6	Universality
7	Nesting
8	Counterweight
9	Prior counter-action
10	Prior action
11	Cushion in advance
12	Equipotentiality
13	Inversion
14	Spheroidality
15	Dynamicity
16	Partial or overdone action
17	Moving in a new dimension
18	Mechanical vibration
19	Periodic reaction
20	Continuity of a useful action
21	Rushing through
22	Convert harm into benefit
23	Feedback
24	Mediator
25	Self-service
26	Copying
27	Inexpensive, short-lived for expensive durable one
28	Replacement of a mechanical system
29	Pneumatic or hydraulic construction
30	Flexible membranes or thin film
31	Use of porous material
32	Change the colour
33	Homogeneity
34	Rejecting and regenerating parts
35	Transformation of the physical and chemical states of an object
36	Phase transformation
37	Thermal expansion
38	Use strong oxidisers
39	Invert environment
40	Composite materials

4. First to match parts and later mismatch parts (to gain advantage), e.g. pocket-knife with one blade, then many blades, finally with scissors, screw-drivers, can openers etc. (Swiss Army Knife).

5. Increasing complexity followed by simplicity through integration, e.g. PCB with lot of components leading to Integrated Circuit.

6. Transition from macro-system to micro-system, e.g. rolled glass sheets to float glass; steel rollers with reducing diameters ultimately lead to molecules of molten tin acting as rollers.
7. Technology follows increasing dynamism and controllability, e.g. wooden pointer, to telescopic pointer, to laser pointer.
8. Decreasing human involvement with increasing automation, e.g. all on-board controls on Satellite.

The underlying guiding principles behind this evolution were that "every system evolves towards increasing ideality" and "evolution continues at the expense of system's own resources". (Note: Contemporary TRIZ software has an *Evolution Trends database* containing over 20 trends and 200 lines of evolution with examples from different processes and products.)

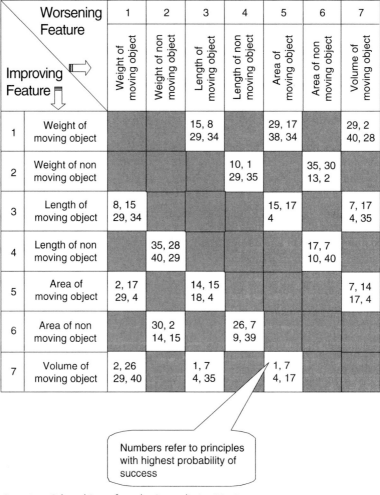

Improving Feature \ Worsening Feature	1 Weight of moving object	2 Weight of non moving object	3 Length of moving object	4 Length of non moving object	5 Area of moving object	6 Area of non moving object	7 Volume of moving object
1 Weight of moving object			15, 8 29, 34		29, 17 38, 34		29, 2 40, 28
2 Weight of non moving object				10, 1 29, 35		35, 30 13, 2	
3 Length of moving object	8, 15 29, 34				15, 17 4		7, 17 4, 35
4 Length of non moving object		35, 28 40, 29				17, 7 10, 40	
5 Area of moving object	2, 17 29, 4		14, 15 18, 4				7, 14 17, 4
6 Area of non moving object		30, 2 14, 15		26, 7 9, 39			
7 Volume of moving object	2, 26 29, 40		1, 7 4, 35		1, 7 4, 17		

Numbers refer to principles with highest probability of success

Figure C7. Selected Rows from the Contradiction Matrix

Whilst TRIZ has been used most effectively in the engineering fields it obviously has applications in many other industries, especially in the chemical and process engineering fields, where significant advances have been claimed in improving yields, product design and resource utilisation [C-21, C-22].

1.8.9
Creating and Robert Fritz

Problem solving is not creating.

Robert Fritz has developed a quite different approach, believing that you need to change your view of the world from one of *problem solving* to one of *creating* [C-23, C-24]. When you are solving a problem, you are taking action to have something go away: *the problem*. When you are creating, you are taking action to have something come into being: *the creation*. However, this relies on having a desire to create and a vision of what you want to create. The power of this vision is in identifying your current concerns and problems, and recognizing that you cannot solve them from your current reality, which allows you to create the future you want.

There are five *common steps* in the creative process, which are guidelines for creating your desired future. These guidelines are for use with the individual, but can then be developed to share visions within teams or even across whole organisations.

1. *Conceive the result you want to create*
   Have an idea of what you want to create. If you could have exactly the result you want, what would that result look like? Write this in the present tense, as if you already have it.
2. *Know what currently exists*
   Look closely at what currently exists in your situation. What does the situation look like? How are you and other people behaving in this situation? How long has this situation persisted? What have you and other people done to change the situation, and what has happened? Be careful not to spend too much time on this. It can be easy to talk yourself out of taking any action.
3. *Take action*
   Because you now know what you want to create, opportunities begin to present themselves. Take advantage of these opportunities. Take action. One caution that Fritz states is that "some of the actions you take will help you move directly to the result you want, but most will not. The art of creating is often found in your ability to adjust or correct what you have done so far." Be prepared to take many actions to reach your desired result. Be prepared to learn from your actions and to adjust your next action, based on what you have learned. Always keep a clear picture of the result that you want to create, and don't give up.
4. *Learn the rhythms of the creative process*
   These are germination, assimilation, completion. At first you'll be excited. Then you'll notice that the thrill is gone. This is when you need to focus on your result, and take action towards that result. Eventually, you will have created your desired result.

5. *Create momentum*

Don't stop now. You get better at creating only by continuing to create. The more you create your desired results, the easier it becomes to create the next result. Fritz uses Mozart as an example of creating momentum. "The more music he wrote, the more he was able to write."

1.9

The Evaluation of the Ideas Generated

Select rather than reject the ideas generated.

After sessions involving any or all of the above methods the ideas that have been generated will need to be evaluated. If the sessions have been productive this is no mean task. It is very time consuming and the process of evaluation needs to be carefully thought through before starting.

When looking at the list of ideas, away from the exciting atmosphere of the generating session, it is easy to feel somewhat disillusioned. It is therefore important to remember that these ideas are only at the germination stage.

- They were not intended to be fully developed ideas.
- Many will have served their purpose in generating more useful ones.
- The promising leads will no doubt need follow up work before they can be completely assessed.

It is easy to prejudge or dismiss ideas in the process of reducing the extensive list to one of manageable proportions. Any evaluation team's job is to select rather than reject the ideas generated. The composition of the evaluation team should lean towards the entrepreneurial and creative types (see Section C, 1.5.1). The objective must be to act as gardeners, keeping fledgling ideas in the "greenhouse" until they are strong enough to be put into the cold [C-25].

A common experience during the process of evaluation is to see that some ideas, which at first sight cannot work, when combined with others do offer distinct possibilities.

The procedure to be adopted during the evaluation phase is as follows.

- Select the criteria by which idea is to be judged. The questions asked might well be; can it be exploited by the company, is it within our technical competence, are the cost and time implications satisfactory, are there any health and safety implications or adverse environmental impact, etc.
- If the list is very long then make a sub group of similar or related ideas that satisfy most of the criteria.
- Select the best from each group. Look at positive features rather than the negative ones. Try to modify the ideas during evaluation.
- Develop the selected ideas in the form of practical proposals.
- Test those designed for *immediate* R&D work by checking how many ways they could fail, a reverse brainstorm.

After a session, it is always a good point to feedback the findings to those who took part in the idea generating process. By so doing commitment for any future activities is more likely and also there will be greater support to the outcome of the study.

One survey, of scientists who use creative thinking tools in their jobs, showed that by far the most commonly used techniques were those of the brainstorming type, and they are also held in highest regard. Morphological analysis methods were much less widely used but were well regarded by those who had actually used them in their job. Synectics and metaphorical analogy methods were little used and were thought only to be of medium value, probably indicating the difficulty of using such techniques correctly [C-1].

In a more recent study by Sony Ericsson of people working in the IT and biotechnology industries, it was found that 81% of people say they have their best ideas out of the office [C-48]. The respondents also said that their best, implementable ideas were had in the following contexts.

25% when socialising
18% at bedtime
6% in the lavatory
4% in the pub

The same survey found that although 65% thought they were creative at their office desks, a higher figure, more than 80%, thought that off-site meetings, especially brainstorming sessions enhanced their creative thinking.

2
The Protection of Intellectual Property

Intellectual property arising from R&D is an important part of a company's knowledge asset and needs protection.

The underlying objective of R&D work is to generate novel information, or intellectual property as it is known in legal terms, in the form of new products, processes, systems and services. The company does not generate intellectual property in industry for the primary purpose of furthering the science, as would be the case for a large part of academic research, but for commercial exploitation. Intellectual property arising from industrial R&D is therefore designed to provide the company with a competitive advantage in the marketplace; it is an important part of its *knowledge asset*. This information will have taken a great deal of time and effort to accumulate and is often the main source for differentiation of one company from another in the same field. It is therefore extremely valuable and needs to be protected from falling into the hands of commercial competitors.

There are three main types of intellectual property that are relevant to the R&D Manager in a chemical or pharmaceutical company. These are inventions in the form of Patents, the company Trademarks, and Trade Secrets, often called *Know-How*.

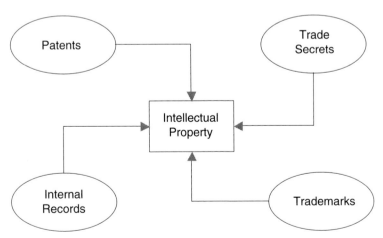

Figure C8. A Company's Intellectual Assets

Research and Development in the Chemical and Pharmaceutical Industry, Third Edition. Peter Bamfield
Copyright © 2006 WILEY-VCH Verlag GmbH & Co. KGaA, Weinheim
ISBN: 3-527-31775-9

There are also internal records of work done and information gathered, which could be of great value to a competitor. These are shown schematically in Figure C8. There is also Copyright, but this is usually of little relevance in chemical R&D.

Most major chemical companies will employ their own intellectual property personnel. These are professionals, educated in patent law, who provide all the support necessary for R&D Managers and their staff. In smaller companies an outside patent agent or legal attorney will be used. Whichever is the case, it is essential that R&D Managers have a working knowledge of intellectual property law. Ensuring adequate protection of inventions will be one of the main responsibilities in an R&D Manager's job description. The level of understanding of the law should be sufficient to provide guidance to the scientific staff and to be able to follow and question any advice given by the legal professionals. Intellectual property law varies slightly from one country to another and the fine details and implications of such differences are best left to the professionals. In this section only the main outlines and common features of the international law will be dealt with, but this will form a good enough base on which a new Manager can build for the future [C-26]. The UK patent law will be used as a model with differences in other countries pointed out as appropriate [C-27].

2.1
Patents

A patent is granted to the inventors, by the relevant government of the country in which an application is made, in exchange for the authors providing a full and complete disclosure of how best the invention may be achieved or operated. Patents, which are passive-protection rights limited in time and geography, may be granted to any number of people. Patents do not in themselves allow the inventor to practice the invention, but prevent other parties from making, selling or using anything covered by the patent in that country for a specified period of time. This time period is almost universally 20 years from the issuing date of the patent.

Patents may be obtained in the following areas.

- Functional articles, equipment, instruments, structures, buildings.
- Chemicals, chemical formulations, and increasingly pharmaceuticals, including synthesis processes.
- Processes/methods for making things or for testing, controlling or treating things. (Biological processes for making animals or plants face moral objections to patenting.)
- Microbiological processes for making or treating non-living things and the products when so made.
- Microorganisms and propagating materials, and research 'tools'.

In the Chemical and Pharmaceutical Industries it is most important to obtain a patent covering inventions on new chemical entities, sometimes called "*compositions of matter*". These composition of matter patents are of great value as they will dominate any future patenting activity by competitors, for instance on materials showing improvements but falling within the generic structure defined in the original patent. They are therefore a potential source of income to the company from licence fees (see Section C, 3.3). Composition of matter patents is relatively easy to protect from misuse by unscrupulous competitors (see Section C, 2.1.3).

Patent protection for novel processes, application methods or machinery can also be obtained. Unless of major commercial significance, the value of filing an application for chemical processes, is often questionable. Chemical manufacturing processes are carried out in a confidential manner with as few details as possible being made public. It is therefore difficult to know if somebody is infringing a process patent in the myriad of chemical plants operating throughout the world. In such circumstances it is often better to treat the process details as trade secrets, relying on a good internal secrecy system to keep these details from falling into the wrong hands (see Section C, 2.3).

Increasingly, in the biochemical, biotechnological and pharmaceutical areas of the industry, work is being carried out in controversial areas such as genetic modification and generation of new entites. Following recent legal judgements, the USPTO has broadened its definition of what can be considered patentable subject matter in the pharmaceutical and biotechnology fields [C-49].

Provided that,

- the claimed invention is new, useful and not obvious
- the specification contains a proper written description that describes the best mode for putting the invention into practice that is known to the applicant
- the specification provides proper enablement to put the invention into practice,

The following can be claimed.
1. New chemical entities, including new intermediates and in appropriate cases new salts, enantiomers and polymorphs.
2. Methods of making new compounds.
3. Methods of treatment, diagnosis etc using new compounds.
4. New methods of making both old and new compounds.
5. New compositions, possibly containing known compounds but in different dosage amounts or forms.
6. New methods of treatment, diagnosis etc. using old or new compounds.
7. New methods of modulating biochemical processes which are carried out in a human, animal or plant.
8. New kits for example containing a new combination of materials or of materials and equipment used for diagnosis or treatment.
9. Newly identified DNA as long as it is claimed in a form which differs from that in which it occurs naturally (that is to say it is claimed in purified or isolated form). Newly created DNA is treated as a new chemical compound.
10. New organisms and parts of organisms such as seeds, for example those containing modified DNA.
11. New vaccines.
12. New vectors, such as plasmids, new hybridomas and new antibodies.

13. New research techniques and in some cases at least the products obtained from using these techniques

This is a highly complex topic for patent activity, one that is subject to change and differing interpretations in different Patent Offices around the world, and expert guidance is needed, that will no doubt be provided to R&D by the company.

Examination of a published Patent Specification will reveal several common features, from whatever the country of origin.

- *Patent Number and Country of Origin and Patent Date* are at the top.
- *Title.* A brief descriptive title containing the keyword descriptors.
- *Inventors.* The names of the inventors and often their companies and addresses.
- *Filing or Priority. Date* and a serial number.
- *Text.* The text contains a detailed description of the invention.
- *Examples.* These are examples of the products or processes given in sufficient detail that "those skilled in the art" can reproduce them.
- *Claims.* These are listed at the conclusion. The most important of these is Claim 1 which should cover the main elements of the invention.

2.1.1
Inventorship

Whilst anybody can apply for a patent it will only be granted, certainly in the UK, to three classes of people.

- The inventor(s)
- Those entitled by law or agreement to preference over the inventor(s), usually the employer
- Successors in title to either of the above

It is therefore necessary to be quite clear who are the inventors of the subject matter of a patent.

During the course of a research project several people will be involved in, or will make some contribution to its success. For instance, these contributions can be in the design of the materials or experiments, the carrying out of the experimental work, the testing of the materials or processes, detailed analytical work, expert consultation or general project team or peer group discussions.

This "team" approach to R&D work can present problems when it comes to deciding who are the inventors to be included on a patent application. It is absolutely essential that the real inventors, and only those, are named on the patent. It is unacceptable to put a person's name on a patent for spurious reasons such as; "she did some good experimental work, it is a just reward" or "his analytical skills helped to prove the structure" or even worse "to keep the boss happy!"

The definition of a true inventor is " one who contributes to a novel feature of the invention as described in the claims" [C-26]. The decision on who are the true inventors is best left until after an independent view from the patent agent has been obtained, on the basis of the evidence provided at the time of generating the application.

Another important feature for the inventor is to keep detailed records of all experimental work. This requirement should have been built into the R&D group's quality procedures as outlined in Section B, 3.4.1.

The most common format for recording experimental evidence or data is the hardback, bound notebook with hand written signatures. The electronic notebook is becoming acceptable to some regulatory authorities but some legal authorities still consider it to be easier to fraudulently edit. More details are given elsewhere in this book (see Section B, 3.3.2.6). The strict compilation of all data in folders for each Project is another excellent method of keeping track on written records, useful in support of inventorship claims and other legal considerations. All such data should be signed and dated at the time of its production.

2.1.2
Filing and Prosecution of the Application

The procedure for acquiring a patent is broadly similar in most countries. In the UK the procedure has the following steps [C-27].

1. Filing of the application
2. Preliminary examination and search of earlier patents
3. Publication of the application
4. Full examination of the application
5. Grant

The application for a patent must contain the request for the grant of a patent, the specification, the claim(s), the abstract and the filing fee.

The timing of filing the application is crucial, as it will set the priority date for all subsequent actions. A Manager must be convinced that there is sufficient information at this stage to draft the specification and to ensure that the claims are valid but also broad enough to capture all the ramifications of the invention. An error at this stage can allow competitors in at a later date. However, too long a delay can lead to the invention, or one that dominates it, being filed by another company. This happens surprisingly often and is most frustrating for all concerned. It is worth remembering that there are many R&D teams out there working hard in very similar areas. It is therefore very much a balancing act between timing and coverage. On balance it is no doubt better to have a patent that has incomplete coverage of the opportunities than none at all or, disastrously, for it to be in the hands of the competitors. Once again an experienced patent officer or agent will provide valuable guidance on all these aspects.

In most companies the process for filing a patent application involves the filling in of inventorship forms by the inventors. There will normally be two forms.

1. *Technical*

This contains the following information for the legal drafting of the patent.

- Details of the Invention
- How it arose
- Who was involved
- What is the prior art

2. *Commercial*

The following information is needed to justify the expense of filing and prosecuting a patent application.

- Which business is involved
- Market potential
- Other commercial opportunities, e.g. licensing
- Breadth of Filing

The question of inventorship and the generation of the invention required for this Form have already been discussed in Section C, 2.1.1. Another major issue on this Form is that of the prior art.

Prior art is that information which was available in the public domain at the time of the production of the data on the invention. The search for prior art is therefore not just the relatively simple matter of checking the literature for conflicting patents. It also involves a detailed search of the literature for any piece of information that discloses any aspect of the invention. Such disclosures include not only patents but also any printed article or oral presentation, sales literature, offers for sale or actual sale of materials, which have been made publicly available. This is a time consuming activity but one which is necessary since it can avoid spurious applications being made which are later rejected by the Patent Office. Additionally, under the law the applicant has a duty to disclose all prior art that can have reasonably been known to exist.

It should be clear from this list of prior art that it is possible for a company's employees to accidentally disclose information, which can be cited later against its own patent. It is imperative that discussions on the nature of the invention do not take place in public before the application has been filed. Disclosures which must not be made include the presentation of papers at public conferences, scientific papers in journals or press releases. Additionally products covered by composition of matter patents must not be offered for sale or test or disclosed in sales literature. For process patents, products made by a new process must not be offered for sale until after the filing date. Similarly, new application methods for a product must not be used commercially until that date. It should also be remembered that there are no international boundaries; disclosure in one country can bar a company from getting a patent later in another country. However, the USA disregards foreign unpublished knowledge and foreign public use.

The filing of the application follows one or all of the following three routes.

1. Individual filings in each of the countries where protection is thought to be necessary by the inventors.
2. European application to the European Patent Office (EPO), designating the member states where protection is required.

3. An international application, or patent corporation treaty (PCT), which leads to parallel national patents in the countries of choice. Most countries are parties to the PCT.

The filing and prosecution of a patent is an expensive activity, the costs rising from route 1 to 3. The costs become particularly high for an extensive filing in many individual countries. In most companies operating at a multinational level filing in several countries is done simultaneously. Hence the importance of a commercial consideration at an early stage, especially if budgets are limited. The patent officer or agent will act on the advice of both R&D and Marketing on the breadth of the filings. The main criterion is whether the invention provides the company with a competitive edge. This competitive edge is generally derived from the direct exploitation of a product or process. However, commercial benefit can also be obtained by indirect means such as licensing. These aspects are covered in Section C, 3.3 and 3.4.

On filing the patent application a priority date is obtained. This date will be cited in all future references to the filing, for instance when conflicting applications appear from competitors, not only in the country of initial application but elsewhere.

There is a time period of twelve months between the initial filing of the application, the priority date, and when the request for examination, either preliminary or substantive, needs to be made to the Patent Office. It is during this period that the work necessary for the filing of the complete patent specification is done by R&D. The work will involve carrying out the further experimental work that will be used to exemplify the claims within the patent application. A considerable amount of work must be done if the patent is to be a strong one. The R&D Manager will need to ensure that sufficient experimental support effort is available to perform this exemplification work. The need to supply this effort is all too easily forgotten in the rush to get on with the product development work until, suddenly, there is insufficient effort or time available for the work to be done in a satisfactory manner.

During the prosecution period it is possible to refine the main claims of the application. It is not possible to materially change or extend them beyond the original boundaries. If it is found that the original claims are insufficient to do justice to the invention, it is possible to abandon and resubmit a new application. This new filing will, of course, have a later priority date. This action will increase the danger of being beaten by the competition, if they have filed an application between the two dates.

The publication of the patent occurs automatically at the end of an 18-month period following the date of filing or priority application. It is important to remember that this publication now constitutes prior art. If the publication would interfere with more important patent applications, or is incomplete it is possible to abandon prior to publication with a consequent loss of the priority date.

If the examiner decides that the application does not contain a patentable invention, usually because there is prior art, which makes the invention obvious, a letter will be written to the inventors informing them that this is the case. In these cases the R&D Manager will be required by the company to ask the inventors to provide evidence that will overcome the objection, or to suggest modifications to the application, which will make it acceptable to the examiner.

Once any objections are overcome by the inventors, and the examiner is satisfied that it is a patentable invention, a patent will be granted after payment of the required fee.

The EPO system also has a provision for anybody to apply for a patent to be revoked, provided this application is made within 9 months of the grant of a European Patent. If this is successful the patent rights are revoked in all member states.

When two or more pending applications have been submitted at around the same date by different parties it is possible in the US to have an interference contest. This involves detailed examination of the written records to see who was actually the first to make the invention. Hence, this is another example for the need of concrete and verifiable records of the work as it was done.

2.1.3
Infringement of Patents

The objective of having a patent is to stop other companies benefiting from the inventions. When a company does try to benefit, by making and selling a chemical product or operating a process for commercial gain, which is covered by another person's patent, then they are guilty of infringement.

Infringement can only occur in the country or countries where a patent exists. This emphasises the importance of the advice given by both R&D and Marketing to the patent agent on the breadth of the filing to be pursued. An R&D Manager should set out to gain a wide knowledge of the countries where the main competitors operate, both in terms of manufacturing and sales, so that the advice can be given with confidence. It is commercial nonsense to suggest the filing of a patent application, in a country where the costs would be hardly covered by the profit from the sales income on a single or range of products.

Once an infringement has been found it is important that the company's patent and legal representatives quickly follow this up, a failure to do so could be interpreted as approval.

The converse of this situation also applies. The company should have a system in place, usually assigned to R&D, which thoroughly checks its own proposed new products and processes, to ensure that it will not infringe somebody else's intellectual property once they are commercialised. If there are conflicting patent rights then a licence to operate will need to be sought from the inventors.

2.1.4
US Differences from the EPO

Because the US is such an important market it is well worth listing separately the special conditions that apply to their patents versus the EPO.

1. *Inventors.* All US patents must be filed in the names of the inventors. Assignment to the company can subsequently be done.

2. *Prior Art.* For the EPO this is anything made public by anyone, anywhere. In the US you can still file a patent up to one year after public disclosure by the inventor. All known prior art must be disclosed; failure to do so will render the patent invalid.

3. *Competing Applications.* For the EPO patent rights belong to the first to file. In the US it is the first to invent, decided by assessment in *interference proceedings*. Hence the importance of keeping accurate dated records of work leading to the invention, e.g. laboratory notebooks.

4. *Technical disclosure.* Has to be detailed enough to meet the US *sufficiency requirements*, and the best example must be described.

5. *Publication.* There is only one and that is when the patent is granted.

6. *Filing Procedures.* In the US there is the opportunity to re-file an improved or supplemental application.

2.2
Trademarks

Trademarks are very important to commercial organisations as they are used to differentiate one company's products from another company's in the eyes of the customer. Such trademarks include words, symbols and devices or combinations of these. Trademarks are registered and are protected by law for a significant time period depending on the country involved.

It is the duty of the Manager to ensure that trademarks are not misused and so lose their value to the company. Everybody is aware that trademarks can start to be used as a noun, for instance Hoover became synonymous with household dust cleaners some time ago. Once this happens then their use as a trademark is called into question. It is easy to fall into this trap in verbal presentations at scientific meetings and also in written papers in the scientific and trade press. The Manager should be on the look out for any signs of this happening, ensure that all staff know the rules, and that a checking procedure for outside publications is in place.

2.3
Trade Secrets

The most difficult intellectual property to protect is that of a trade secret, or *"know how"*, which is specific to that business. Know-how is often immediately exploitable information, for instance manufacturing process details or the constitution of mixtures, which when obtained avoids the need to carry out the more time consuming and expensive aspects of R&D.

Contracts of employment normally contain a clause stating that when a person leaves the employ of that company and is recruited by or works with a competitor then that person will not divulge any trade secrets to the competitor. This is enforceable by law in a many countries.

Several high profile cases in recent years have shown just how difficult this is to prove and police by the injured party. The intellectual learning gained whilst in one R&D department in Company A is especially difficult to monitor when a person moves to Company B. One can only rely on the professional attitude of the employee whilst taking due precautions over very sensitive information. These precautions include keeping a tight hold on secure documents and restricting access to the "Company Secret" level information (see Section C, 2.4).

In the case where the filing of a patent is impossible or it would be very weak, know-how is all that is left and it is very valuable to the company. A good example is provided by those companies who specialise in the formulation of known compounds for specific needs, especially in the speciality chemicals sector. Some companies in this area have been known to restrict the access of each person, especially those in R&D or technical service, to a specific segment of the business. This is done to try and ensure that all the know-how does not go out of the company, when one employee leaves to join a competitor or to start a company.

2.4
Internal Records

Several times in the preceding items the need to record and securely store information relating to intellectual property has been mentioned. The more important information produced by R&D is recorded in several ways, the most common being the following.

- Laboratory notebooks, both hard copy and electronic, for experimental work
- Results from the application testing of materials
- Analytical results and information on chemical structures (LIMS)
- Project folders, containing data on individual projects produced by the team
- Technical reports, interim or on completed projects
- Manufacturing processes, including plant diagrams
- Patent specifications

These are produced in both hard copy format for physical archiving and in machine-readable form for storage in computerised databases. In either case they must be professionally managed, properly archived and backed up and secure from theft. In the larger companies this will be done by competent staff or by using a contract company. This is unlikely to be the case in the smaller companies and an R&D Manager will probably have to ensure the security of this information as part of the job. Employees cannot be expected to protect the organisation's secrets if they are given no guidance by way of an appropriate security classification of documents and computer records. And, if everything is "secret", nothing is! The right culture in an organisation that shows due respect and regard for the competitive value of its knowledge is most important.

Information can be categorised in three ways.

1. Information which is general and of little or no commercial value.
2. Information which if it fell into the hands of a competitor would be detrimental to the company.
3. Information which if it became available to the competition would cause serious damage to the company's business.

The Manager should put all information produced by the R&D Group into one of these categories. All information, in whatever format, should be stored safely and securely but especially that of Categories 2 and 3 described above. The R&D staff will want access to this material from time to time but this must be controlled. Category 1 is made available to all, Category 2 only to those who need it for their job, with Category 3 restricted to those with an essential need to know, and then only with a senior Manager's approval.

Some of this may seem unnecessarily restrictive, but it is possible to walk out of a company with an enormous amount of information on a 3.5" floppy disc, CD or other storage media using an unattended PC connected to an insecure database.

2.5
Code of Best Practices for Handling IP

Based on the previous discussion a code of best practices for ensuring protection of a company's intellectual assets can be drawn up [C-28].

1. Treat all technical developments that signpost a new product or process or improve on existing ones as patentable.
2. Keep patentable developments confidential at least until a patent is filed. There should be no pre-filing public use or disclosure.
3. Keep competent R&D notebooks. Have them read, dated and signed whenever something new and significant is reported.
4. Patents, when issued, stop competitors doing what is patented, where patented (technical exclusion + market exclusion). The challenge comes in drafting, demonstrating and justifying broad patent 'claims' (which define what is patented). Initial findings start this process; much work is needed to finish it.
5. Technical Service/Sales/Marketing co-operation during the patenting process is essential to avoid prejudicial early releases.
6. Invention does not stop at the factory gate or with delivery of product. Novel uses and innovative applications are also patentable.
7. You cannot patent something already known or published anywhere (but the USA disregards foreign use/unprinted disclosures). This is the "prior art".
8. The internal stock of knowledge is not the same as prior art, but some/much will be. A patentable invention is one that is new and non-obvious over the prior art.
9. The early publication of patent filings (18 months from first filing) gives both you and competitors an early awareness of each other's technical targets/interests—and potential patent blocks. Be as well informed as they will be!
10. Improvement patents can turn a technical promise into a commercial triumph.

They have added competitive merit if you get them and great nuisance value if your competitors get them first. Use your lead-time effectively.

11. A patent portfolio covering a group of related inventions is stronger than the sum of the individual patents.

12. Keep good records of what you make, how you make it and how you use, install and deliver it. But secret use in the USA gives you no priority over a later US patent. Elsewhere actual local use *may* create a "right to continue to work".

13. Freedom of operation against later patenting by others is assured by publication but not by patenting as such. Publication without patenting creates a level playing field everywhere.

14. Patenting only in some of the countries where the invention has commercial relevance makes an automatic gift of the invention to your competitors in the others.

15. Patents cannot prevent R&D evaluation of the invention by competitors but the 'crucible' in which the know-how critical for successful commercial use is developed (and in which the experience that springboards yet further enhancements is acquired) is so often commercial exploitation. Deny your competitors that in their markets and you sustain competitive advantage.

16. Patenting is expensive but more so is the cost of designing around your patents to create functional equivalents if your competitors are in catch-up mode.

17. New technologies flow from technical creativity. Given that, patents sustain business advantage; nothing else in the business tool-kit is as effective or as enduring.

18. When you are constrained to work co-operatively with third parties to develop or prove new technology do it on a contractual basis that stipulates publication rights, patenting rights and respective exploitation and licensing rights.

19. Patent strategy is a business plan aided by expert patent opinion/advice. Patenting is the joint endeavour of the inventors and patent attorneys working to a commonly understood objective consistent with the patent strategy.

20. Don't close your mind to cross-licensing patents.

3
The Exploitation of Opportunities

Develop IP in house, joint venture, licence or file for the future.

Having developed some intellectual property to the level where it is open to exploitation a decision must be made about how this should be done. R&D and marketing personnel will be involved in an evaluation of the product or process to see if it is suitable for development within the company. If, for a variety of reasons, the answer is no, exploitation by other means will be considered. One way could be in a *joint venture* with an appropriate partner. Alternatively, it is possible to *license* or sell the intellectual property to another company, if this fits in with the company strategy. The final option is to abandon the project at this stage and to leave the data in the company secret files, for reconsideration at some future date. These options are shown in Figure C9.

The opportunities for exploitation, described above, arise from developments within the company and any transfer of technology would be outward. There are also those opportunities which arise from outside the company, but that could be developed within the company, if there was an inward transfer of technology. The transfer of technology is an increasingly common way of exploiting the opportunities from

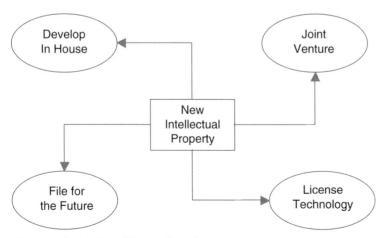

Figure C9. Exploitation of New Intellectual Property

Research and Development in the Chemical and Pharmaceutical Industry, Third Edition. Peter Bamfield
Copyright © 2006 WILEY-VCH Verlag GmbH & Co. KGaA, Weinheim
ISBN: 3-527-31775-9

R&D on a wider basis between companies. In this book we will discuss *technology transfer* mainly on the basis of its inward movement to the company.

3.1
Development within the Company

This is by far the most common route for exploitation of the results from R&D as, rather obviously, the majority of its work is targeted towards achieving objectives set by the current business.

The document listing the R&D target will have laid down criteria that a new product or process must meet if it is to go forward to commercialisation. Typically the following questions must be answered about projects.

- *Quality.* Does the new product meet the technical specifications required by the market, does the manufacturing process consistently produce material suitable for all outlets, does the application process produce the desired effect in all cases?
- *Cost.* Does the new product meet the profitability criteria, does the new process make the required savings, is the application process cost effective?
- *Capital.* Can the new product be made in existing plant, can the savings from the new manufacturing process justify the capital required, does the application process fit in with the customers existing machinery?
- *Environment.* Is the environmental impact of the new product or processes improved over the existing ones?
- *Safety.* Is the new product safer to handle, is the process less hazardous?

It is unlikely that all the criteria set in the target document will be met and the decision is often made as a balance of several factors. This aspect of the R&D methodology is covered in greater detail in Section D.

3.2
Joint Venture

Sometimes it is apparent that opportunities arising from the results of R&D work, especially those related to new business ventures or from research of a more strategic nature cannot be exploited effectively by the company on its own. In these cases it may be that the innovation can be introduced to the market by jointly working with another partner. This is the joint venture route to exploitation. Having the right partner is of crucial importance, the wrong choice being the cause of most failures in joint ventures.

The reasons why a company could decide that in house exploitation is not the right route to follow typically include the following.

- High capital requirements
- Inadequate plant
- Outside the core competencies

- Lack of market knowledge
- Sales too small to justify development
- The time is too short

There is often an overlap between one or more of the reasons and this is especially true of high capital requirements and inadequate plant.

> As an example, a new product has arisen from research, which is exciting but is one that would require an investment of capital way beyond the means of the company. The capital required is for the construction of a suitable plant and also for the extra staff needed to develop the product to a state ready for launch into the market.

This example is quite common in smaller companies, especially those whose research is based on a specific technology, such as biotechnology or materials chemistry. If the decision taken by the company is to exploit the technology, rather than license, one way to obtain the necessary capital is to seek funding from a venture capital organisation. This gives the company the opportunity to grow but clearly with some loss of independence, as a stake in the equity is usually the price for the support. Another option is to find a company that is already operating in the area and to suggest joint exploitation of the new product. It is unlikely that a Research Manager in a medium to large company would initiate such contacts; this is normally done by Marketing or a New Business Manager. R&D will be expected to supply the necessary technical input to the negotiating team. In small companies the role of R&D is much more integral in the process.

> In another example, a new catalyst has been discovered which is ideal for use in the catalytic hydrogenation of nitrohydrocarbons by a continuous process. It has an enormous potential for increasing the business of the company. Unfortunately, the current plant was designed to handle batch processes and is unsuitable for modification. The generation of capital to construct the plant is a problem for the company.

Seeking support from a venture capital company might once again be the answer. However, in this case there are alternatives worth consideration. It is possible to outsource, rather than joint venture, the manufacture to a company who already has a suitable plant, something that is also common in the early stages of new product development (see Section D). Alternatively a joint venture could be arranged with another company, where there is a sharing of the capital investment, in return for an equitable share of the profits from the products made in the plant by the new process technology.

An interesting combination occurs when exploitation is beyond the core competencies of the company and the sales of a particular product would be small. A theoretical example of such an occurrence in the electronic materials for displays area is illustrated below.

> A major electronics company is carrying out research into multi colour pixels for light emitting diodes. R&D has discovered, during random screening, that the

wavelength of the light that is emitted can be varied by doping the normal inorganic solid-state materials with an organic compound. Variations on the structures of the organic compounds, which produce this effect, are difficult for them to make, as knowledge of the synthetic methodology is outside their core skills. They find, by searching the literature that similar compounds are used in the protection of a variety of metals from corrosion and are sold for such outlets by a number of speciality chemical companies. They approach one such company to see if they will supply a range of both existing and new materials for test in this new outlet. The speciality chemical company discovers that, even if they are successful in providing materials that meet the target, the level at which the dopants are used would mean that the income from the sales of the chemicals could not justify spending the money on the research. The electronics company accepts this and an agreement is struck whereby the research costs of the speciality company will be covered up to the point when a decision is made on the development of a particular product. If the decision is to proceed then the chemical company will receive a percentage of the sales of the device instead of payment for the chemicals supplied. Joint research is then carried out, the necessary joint patents filed; the device is developed and successfully launched. Everybody is happy.

These different cases illustrate that the exploitation of the results of R&D work by forming joint ventures is a valuable way forward for a company. The joint venture and partnership mode of working is an increasingly attractive option for companies to follow especially since the globalisation of the markets and companies has increased.

The successful outcome from a joint venture in a technological development is not guaranteed but there are ways in which the path can be made smooth. A study of a range of technology partnerships or joint ventures has identified ten factors that appear to be critical to the success of such ventures [C-29, C-30]. Five of these apply during the setting up the relationship and the other five operate during the relationship. These are eminently sensible and simple rules and are ones that fit in with the experience of the author when working within such joint ventures.

Setting up the relationship
- Senior managers should define clearly the goals tied to each partner's business strategy.
- Each partner must understand the other's strengths and weaknesses.
- Before commencing the work all staff must be made aware of the value of the partnership to the company.
- Ownership of the intellectual property must be defined clearly.
- The legal agreement must contain fair and equitable exit clauses.

During the relationship
- The process must be continuously managed.
- The partnership is between people, good relationships are critical to its success and must be maintained.

- Loss of a key person can be traumatic, confidence in the relationship is needed to deal with this if it happens.
- Each partner needs to adapt to the other's operating style.
- Senior management must show commitment and involvement at key steps.

3.3
External Licensing and Assignment of IP

The external licensing of intellectual property is very valuable in the case where a company wishes to retain the patent but also wants to exploit the knowledge through other parties. The licence is usually granted in return for money, this can be in the form of a lump sum but more often it is in return for a percentage of the sales income which is generated by the other partner. If the licence is granted to a single party it is called an *exclusive licence*, if to more than one it is *non-exclusive*. A typical example of where licensing is appropriate is as follows.

> The patent in question covers a class of biocides that are being marketed by Company A. Company B approaches Company A, the owner of the patent, in order to seek a licence to make and sell a compound, the structure of which is dominated by this patent. The decision to grant a licence is a purely commercial one. In this case the financial benefits to Company A are seen to outweigh the disadvantages. A licence is granted to Company B in return for 5 % of the net sales income for the remaining lifetime of the patent.

The accidental discovery of products or processes by R&D that are outside the core competencies of the company does happen. Often this is when it is least expected and exploitation on a reasonable time frame is impossible because it was unplanned. In these cases the assignment of a patent to another company much better equipped to develop the invention is a good option.

> Let us consider the case where chemicals are being synthesised to meet a particular therapeutic target within a pharmaceutical company. The products fail to satisfy the desired target, but a diligent research chemist notes that there are certain structural similarities in some of the molecules that are in common with chemicals that have been reported in the scientific press to be excellent insecticides. These common features suggest they may have similar modes of action when used as insecticides. The researcher discusses this observation with the Research Manager who agrees that they do offer a real chance of working in this area. The Research Manager agrees to make limited funds available for the chemicals to be tested at a local independent research institute. The compounds prove to be very effective insecticides and a patent application is filed. The company has no involvement in the agrochemicals business and therefore cannot directly benefit from this finding. Negotiations are held with an agrochemicals company. Samples are released to this company under a secrecy agreement and the test results confirmed. The patent is then assigned to the other company in exchange for a sizeable sum of money.

3.4
Inward Technology Transfer

There is an increasing need for businesses to bring technological developments to the market place as quickly as possible. The concept of core competencies and the skills audit procedure has already been discussed in Section A. The debate over a company's core competencies is usually held at a strategic level, involving senior management since such competencies cannot be built quickly or cheaply. The skills audit is a more operational level task carried out by middle management. However, both of these methods are of great help in defining the limits and capabilities of the R&D function within the company.

In order to meet a desire for rapid development of technology outside the core competencies of the business it may be necessary to "buy in" some key parts of this technology. This can be done either from another part of the company or more usually from an external source. This is the process of inward technology transfer. R&D Managers are involved in ensuring that this transfer of technology is carried out in a smooth and efficient a way.

A set of factors, from a detailed study of the literature, have been identified as a great help in fostering technology transfer [C-31].

- High quality of incoming information
- A readiness to look outside the firm
- A willingness to share knowledge
- A willingness to take on new knowledge, to licence and enter into joint ventures
- Effective internal communication and co-ordination mechanisms
- A deliberate survey of potential ideas
- Use of management techniques
- An awareness of costs and profits in R&D departments
- Identification of the outcomes of investment decisions
- Good-quality intermediate management
- High status of science and technology on the board of directors
- High quality chief executives

If the transfer is from one commercial organisation to another then it should be relatively straightforward. The agreement with the company should include a clause for the provision of the necessary technical support during the introduction of the technology. If the technology is being transferred from an academic institution, e.g. a university or college, or from an independent research institute it can be more difficult to handle. This is because academic researchers do not, in the main, wish to be involved in the more routine work that is necessary for commercialisation of a technology. In the latter case a company would be well advised not to move too far outside its core technical skills.

3.5
Knowledge Management

Managing knowledge efficiently and innovatively is a requirement in a knowledge based economy.

Over the last few years Knowledge Management (KM) has become one of the hottest topics on company agendas, being seen most importantly, as a way of gaining competitive advantage in knowledge based economy. This is no doubt part due to a backlash against the wave of business reengineering that has been actively pursued by most global industries since the 1980s. The knowledge worker is a buzzword, and the idea of the knowledge company has generated a new role within such companies; the Knowledge Manager.

So what is Knowledge Management? The most favoured definition found in a survey by over 70% of the respondents was as follows [C-32].

> *The collection of processes that govern the creation, dissemination, and utilisation of knowledge to fulfil organisational objectives.*

It is therefore a business focussed activity, and is the process by which a company or organisation manages the knowledge that its employees gather or gain in the operation of its daily business, both inside and outside the organisation. In other words it is about the useful networking of people and their information to the benefit of the organisation, as opposed, for instance, to pointless desktop publishing and intranets. The use of information technologies to integrate, bridge cross-functional boundaries and to provide global networks for knowledge sharing is crucial, the most commonly used for this purpose are online information systems, document management and groupware. One has to be a little careful here for as Drucker argues, key knowledge workers carry inside their heads a company's primary means of production, namely knowledge, and often react against formal information systems [C-33].

The above-mentioned survey also found that there are seven main approaches to KM being used in industry [C-32]. These areas are as follows.

- Knowledge as an intellectual asset
- Knowledge as an HR activity
- Virtual organisations, strategic
- Philosophical
- Technological process

From this list it can be seen, not surprisingly, that knowledge and its management are of great relevance to the R&D function, and in fact R&D leads the way in KM.

The Cranfield School of Management in the UK has developed a model, which charts the relationship of knowledge and its components to business actions and results. The traditional use of the so-called DIKAR model is shown in Figure C10 (a). Starting from the left, value is aggregated by gathering basic data, information, gaining knowledge that leads to "informed actions" which in turn generates business results.

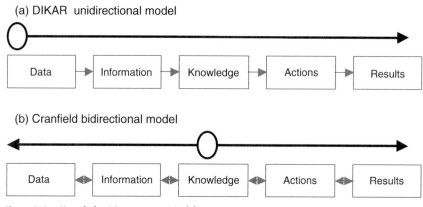

Figure C10. Knowledge Management Models

The approach taken by Cranfield involves beginning from the opposite end. Starting from the desired business results the first step is to decide what actions are necessary to achieve them. The next step is to determine what knowledge is required to perform the actions. The information and data systems are then constructed to support the knowledge-action set. This has the merit that it is business demand led rather than fed by data and technology push. The knowledge manager can use this model to either look upstream or downstream from the knowledge box as shown in Figure C10 (b). Looking downstream from the knowledge box involves very different issues from looking upstream as shown in the example below.

> Looking downstream. A worker at a research establishment has "knowledge about XYZ". This would benefit the Knowledge Manager's company in one of its R&D programmes. This could be the result of work done in the laboratory. This knowledge can be thought of as a body of information, formally written down and capable of being readily digested into the interested company's systems. The issues for KM are identifying the knowledge, its location, validating its value, obtaining it in a useful form at a reasonable cost, determining where it is most useful in the business and making it available there in an appropriate form, using suitable technology, for example with formal learning loops and finally insuring that the knowledge is used beneficially.

3.5.1
Explicit and Tacit Knowledge

In a business context there are two working definitions of knowledge: knowledge as a *"body of information"* and knowledge as *"know-how"*. These are often referred to as *explicit* knowledge and *tacit* knowledge.

The generation of these two types of knowledge in an individual is described in the *Learning Cycle* devised by Kolb in 1974 and illustrated in Figure C11. As can be seen tacit knowledge is attained by acquaintance with a topic and involves divergent think-

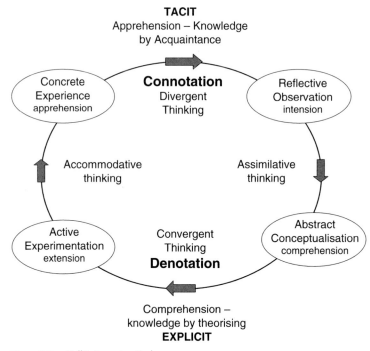

Figure C11. Kolb's Learning Cycle

ing to rationalise that experience. On the other hand explicit knowledge involves comprehension in a convergent thinking mode, leading to a right or wrong answer.

In the technology functions of the chemical industry tacit knowledge is usually applied to that knowledge which is not patentable, the know-how of the organisation, including how to make things happen in your own organisation. These are the tricks of the trade encompassed in process details or formulations used within a company, or knowledge gained by individuals in their daily work and not formalised in any way. It is therefore the information that is passed between research workers in an unstructured manner. It is the most difficult type of knowledge to manage and often disappears out of the front door when long serving workers retire.

3.5.2
Knowledge Repositories

The main way of coping with this large body of information is to lodge it within a repository or database. Davenport has reviewed the different types of knowledge projects operating within knowledge-based companies [C-34]. The majority of projects in his survey fell into what Davenport called a repository or a "bucket-o' knowledge" type. Here the objective is to collect some form of knowledge that has been extracted from the human brains and then stored in a technical system for later access.

- *Type one repository.* The knowledge is held in structured form, usually documents. Hewlett Packard, for example, have systems that store sales oriented documents than can be accessed by their sales forces in selling computers.
- *Type two repository.* The knowledge is less structured and consists of insights and observations of employees, called "discussion databases" or "lesson-learned" systems.
- *Type three repository.* The knowledge is not held in itself but pointers to those who have the knowledge. Hewlett Packard has expert databases for consultation by its researchers in its laboratories.

It is therefore important to know how to access and trawl thorough this internal knowledge so that you can add value to the company, e.g. Chemical structural databases, project data, information from theoretical modelling.

Of great interest, from a technology management perspective, are those companies that are looking at the value of their knowledge-based repositories. One of these, Dow Chemical, focussed on the management of value by harvesting income from little used patent and licence assets. This is dealt with in greater detail in Section C, 3.5.4.

3.5.3
Knowledge Management and the Technological Function

Tacit knowledge of the individual needs to become the explicit knowledge of all.

The R&D function of any organisation generates a vast amount of data and information, produced at great cost, the majority of which is worthy of retention for future use. The more important information is recorded in several ways, the most common being the following.

- Laboratory notebooks for experimental work
- Results from the application testing of materials
- Analytical data
- Chemical structural data
- Technical Reports, both interim and on completed projects
- Project reports or folders, containing data on individual projects produced by the team
- Manufacturing processes
- Product specifications
- Patent Specifications

These are stored in hard copy form and increasingly in machine-readable form in computerised databases. These databases and repositories contain the explicit intellectual capital of a business and ownership can be the source of great argument when part of a business is sold off or merged with another. This information therefore must be professionally managed for the benefit of the business. KM skills also include the knowledge about how to make all this data come together to some desired purpose in including avoiding undesirable out comes, e.g. Understanding HSE issues and their effects on lead times.

As stated, this is essentially explicit knowledge, but what about tacit knowledge, how should this be handled? How can the tacit knowledge of one individual become the explicit knowledge of others?

In their seminal work Nonaka and Takeuchi make the point that explicit and tacit knowledge are not separate entities, they work in tandem and are changed from one to the other by the social interactions occurring during our creative activities [C-35].

They consider that this knowledge conversion occurs through four processes.

1. *Socialisation.* The sharing of experiences, on the job training of "apprentices" by "craftsmen" *(tacit-tacit)*.
2. *Externalisation.* Articulation of tacit knowledge to explicit, e.g. by writing it down, making images, using metaphors or creating models. This is key to the knowledge creation process *(tacit-explicit)*.
3. *Combination.* The process of creating knowledge systems, e.g. documents, meetings, networks and databases, as in the list of recorded information found in R&D as given above *(explicit-explicit.)*
4. *Internalisation.* Converting explicit knowledge into unconscious actions, e.g. the learning by doing so that actions are at the "fingertips" *(explicit-tacit)*.

The biggest challenge to R&D Managers in this area is to get people to share their tacit knowledge with others and hence convert it to the explicit knowledge of the organisation. This is knowledge creation at its most fundamental. Such sharing needs some form of formality, a good example being collecting the knowledge from the team members just when a project is completed. (See also Section D)

All this points to encouraging effective team working within the R&D groups, where each member can bring their personal expertise to the common table. This is best done when this collective knowledge is being applied to a shared problem or as part of a self-organising, multidisciplinary innovation or project team. Nonaka and Takeuchi propose that the team will go through the following processes leading to a leverage of knowledge and innovation in the company; *Sharing Tacit Knowledge, Creating Concepts, Justifying Concepts, Building an Archetype, Cross-levelling of Knowledge.* This process is illustrated in Figure C12.

Over many generations, there has never been any question that a technological function within a business should retain its explicit knowledge in some form of repository. The real question is for what purpose. Usually this is in order to build on the data produced for future projects and to protect the information from leaking out to competitors, most obviously by filing patents but also by maintaining a secret and secure information system.

Potentially there is another type of value to be gained from the information in company databases and other repositories. If the company cannot exploit a project inhouse it has been common practice to license this technology to a competitor for a fee. This was often done in response to an approach from the competitor but Dow Chemical in the 1990s adopted a more structured approach to generating value from its technological knowledge base.

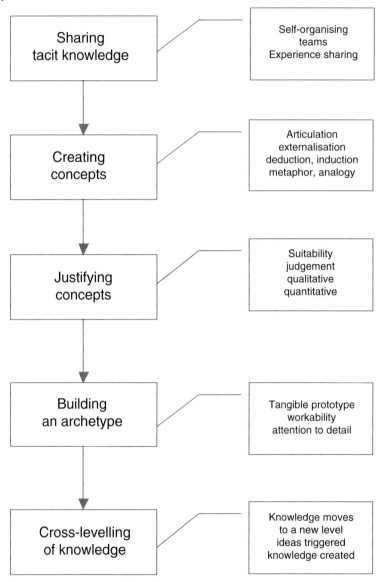

Figure C12. Organisational Knowledge Creation Model (Nonaka-Takeuchi)

3.5.4
The Dow Chemical Project in Knowledge Exploitation

In the early 1990s Dow Chemical considered that it was not exploiting to the full extent its intellectual capital, generated over many years requiring a high-level of expenditure [C-36]. So it set up a small team to examine how it might re-engineer its systems and processes to create more value. The main asset was a disorganised patent portfolio of 29 000 that it was spending $30 million each year on maintaining.

The team needed to measure the value of those assets, and required new processes, tools and advocates to demonstrate how Dow's patents and other knowledge assets could be more effectively used beneficially. Fortunately, they had the backing of senior management who provided an annual budget of about $3 million for a corporate office team to work on these ideas. They were very focussed, sharing the same office space, where they spent a great deal of time creating for the company new processes for Intellectual Asset Management.

Rather than announce some grand plan to manage intellectual capital, the group decided to be make changes of more evolutionary nature and not try to be too revolutionary. So they started with patents, an asset with which many people in the company were familiar.

While the group knew that Dow possessed an array of other intellectual assets (such as know-how, copyrights, trademarks and trade secrets), it reasoned that patents was an area in which it had a high probability of success, could demonstrate obvious value, and implementation of its new processes could be done quickly.

Key to successful development of these new processes was as initial survey of current processes. This survey required all the main stakeholders in the company's businesses to gather in a single room to map out existing patenting activities, roles and relationships. Once this step was completed the group began to consider how it could improve the management of patents in acceptable manner. This effort was critical to achieving buy-in, and engendering trust throughout the corporation for the group's new Intellectual Asset Management Model (IAM).

The IAM Model involves six phases.

1. Strategy
2. Competitive assessment
3. Classification
4. Valuation
5. Investment
6. Portfolio

The group chose to start with the portfolio phase, reasoning that the company already had patents that were under-utilised. Each patent needed to be identified to determine whether it was still active, and find a business that would take over financial responsibility for its maintenance or development.

Subsequently, each business was asked to classify each of its patents in one of three categories: "using", "will use" or "will not use" the patent. The businesses added other designations such as whether the patents were to be licensed or abandoned.

The strategy phase focused on integrating the patent portfolio with business objectives in order to maximise its value. It also identified gaps in the portfolio that needed to be addressed. This phase is connected to the valuation and competitive assessment phases.

In the valuation phase, a value is placed on the asset for licensing, opportunity prioritisation or tax purposes. In association with the consulting firm A.D. Little, Dow developed a comprehensive intellectual property asset valuation tool known as the "Tech Factor Method". The Tech Factor, which relies on several industry-accepted methodologies, facilitates a quick and inexpensive financial valuation of intangible assets within a particular business unit. It enabled Dow to value the monetary contribution of each property or asset as a percentage of the business enterprise's total net present value.

In the competitive assessment phase the knowledge, capabilities and intellectual assets of competitors is determined. The company accomplished this using a well established tool called a "patent tree"; a map of opportunities that incorporates the patents of both Dow and its competitors, evaluating such factors as dominance, breadth of coverage, and opportunity openings. This is in effect creates a "knowledge tree" for looking at such assets as know-how within a strategic context.

In the final investment phase a judgment is made on whether to put more money in R&D, enter a joint venture or license a technology from outside in order to meet business objectives, basing the decision on a prior assessment of the company's knowledge gaps. If the company successfully obtains a needed technology or secures an appropriate patent, the intellectual asset is incorporated into the portfolio and the process repeats itself.

Using this model Dow was able cut its patent tax maintenance obligations by $40 million and other administrative costs by at least $10 million more over ten years. It abandoned (or donated) patents that were no longer of value to the company. The company was able to capitalise on the revenue potential inherent in the patents and estimated that it increased its annual licensing income from $25 million in 1994 to more than $125 million in 1997.

A further benefit from this intellectual asset management was that business people came to realise the value of its patenting activity as well as recognising the opportunities associated with licensing and other related activities. Consequently, they became much more proactive in seeking ways to leverage these technologies outside the firm.

In order to sustain these successes, Dow built a network of people, known as Intellectual Asset Managers, charged with integrating these ideas into the operations of the company's various businesses. In 1997 there were 30 such IAMs, 15 of which held the position on a full-time basis. They were responsible for developing and maintaining an intellectual asset plan aligned with the business strategy; reviewing the IA portfolio at least once a year; and identifying key intellectual assets. They were also responsible for advocating the Intellectual Asset Management vision.

References | 211

References

[C-1] EIRMA Workshop Reports IV. *Stimulating Creativity and Innovation.* EIRMA, Paris 1993.

[C-2] TANNER, D., *Creativity Development in DuPont,* The Nikkan Kogyo Shinbun Ltd., Tokyo, 1998

[C-3] GOUGH, H., THORNE, A., *Portraits of Type: An MBTI Compendium,* Consulting Psychologists Press, Palo Alto, 1991

[C-4] MYERS, I., B., McCAULLEY, M., H., *Manual: A Guide to the Development and Use of the Myers-Briggs Type Indicator,* Consulting Psychologists Press, Palo Alto, 1992

[C-5] DRUCKER, P., *Innovation and Entrepreneurship,* Harper Collins, New York, 1995

[C-6] SMITH, N. I., AINSWORTH, M., *Managing for Innovation,* 2nd edn., Management Books 2000 ltd., Chalford, 1998

[C-7] MAGRATH, A. J., *Planning Review,* 20, **1992**, 12–18

[C-8] ROBERTS, R. M., *Serendipity: Accidental Discoveries in Science,* Wiley, New York, 1989

[C-9] MORRIS, P. J. T., CAMPBELL, W. A., ROBERTS, H.L. (eds.), *Milestones in 150 years of the Chemical Industry,* Royal Society of Chemistry, Cambridge, 1991

[C-10] DE BONO, E., *Lateral Thinking for Management,* Penguin, London, 1971

[C-11] MAJARO, S., *Managing Ideas for Profit; The Creative Gap,* McGraw Hill, New York, 1991

[C-12] Examples include www.cul.co.uk/creative

[C-13] MICHALKO, M., *Thinkertoys,* Ten Speed Press, Berkeley CA, 1991

[C-14] MICHALKO, M., *Thinkpak,* Ten Speed Press, Berkeley CA, 1994

[C-15] CLEGG, B., BIRCH, P., *Instant Creativity,* Kogan Page, London, 1999

[C-16] ROUKES, N., *Design Synectics – Stimulating Creativity in Design,* Davis Publications, Worcester, MS, 1988

[C-17] GORDON, W. *Synectics,* Collier, New York, 1970

[C-18] BUZAN, A., *Mindmap Book,* Dulton, New York, 1996

[C-19] DE BONO, E., *Six Thinking Hats,* Penguin, London, 1990

[C-20] ALTSHULLER, G., *Creativity as an Exact Science,* Gordon and Breach, New York, 1988

[C-21] http://www.triz-journal.com, see also www.mazur.net/triz

[C-22] DOMB, E., *Managing Creativity for Project Success,* Proceedings of the 7th Project Leadership Conference, June 2000

[C-23] FRITZ, R., *Creating,* Fawcett, New York, 1993

[C-24] FRITZ, R., *The Path of Least Resistance for Managers,* Berret-Koehler, California, 1999

[C-25] ALLAN, D., KINGDON, M., MUSIN, K., RUDKIN, D., *?What If!,* Capstone, Oxford, 1999

[C-26] LATHAM, D., BRADBEER, A., *J. Soc. Dyers & Colourists,* 110, **1994**, 58

[C-27] UK Patents Act 2004, The Stationery Office, London. www.patent.gov.uk

[C-28] *IPR Module,* MSc in Chemical Technology and Management, University of Strathclyde, 2000

[C-29] SLOWINSKI, E., *CIMS Report,* LeHigh University, Spring/Summer 1992

[C-30] SLOWINSKI, E., HULL, F., *Research Technology Management,* Nov-Dec, **1990**, 16–20

[C-31] TROTT, P., *Innovation Management and New Product Development,* 2nd edn. Prentice Hall, Harlow, 2002, p. 359

[C-32] Cranfield School of Management, *The Cranfield Information Strategy and Knowledge Survey: Europe's State of the Art in Knowledge Management,* Cranfield, 1998

[C-33] BARKER, A., *The Alchemy of Innovation,* Spiro Press, London, 2002, p. 165

[C-34] DAVENPORT, T., *Secrets of Successful Knowledge Management,* Knowledge Inc, February 1997, http//:welcom.com/quantera/Secrets.html

[C-35] NONAKA, I., TAKEUCHI, H., *The Knowledge-creating Company,* Oxford University Press, Oxford, 1995

[C-36] MANSCO, B., *Dow Capitalizes on Intellectual Assets,* Knowledge Inc, March 1997, http//:welcom.com/quantera/Dow.html

[C-37] DE BONO, E., *Teach Your Child to Think,* Penguin, London, 1993

[C-38] THOMPSON, C., C., *What a Great Idea,* HarperTrade, New York, 1992

[C-39] OSBORN, A., F., *Applied Imagination,* Scribners, New York, 1961

[C-40] OLSEN, R., *The Art of Creative Thinking,* HarperCollins, New York, 1986

[C-41] MORGAN, M., *Creating Workforce Innova-

tion, Business & Professional Publishing, 1993

[C-42] Zwicky, F., *Discovering Invention, Research through the Morphological Approach,* Macmillan, New York, 1969

[C-43] Vance, M., Deacon, D., *Think Out of the Box,* Career Press, 1997

[C-44] Williams, R., Stockmyer, J., *Unleashing the Right Side of the Brain,* Stephen Green, 1987

[C-45] Kosko, B., *Fuzzy Thinking,* Time Warner, New York, 2000

[C-46] McKim, R., *Experiences in Visual Thinking,* Brooks/Cole, 1980

[C-47] Examples include www.cul.co.uk/creative, www.innovationtools.com, www.sixhats.com

[C-48] Milne, J., The Guardian, Business Solutions, 27.01.05, 5

[C-49] Richards, J., *US Patent Law and Practice with Special reference to the Pharmaceutical and Biotechnology Industries,* 2002, www.ladas.com

[C-50] Kirton, M., *Kirton-Adaption-Innovation Theory,* Occupational Research Centre, Newmarket , UK 1988 www.kaicentre.com

[C-51] McHale, J., Training & Development, October 1986

Section D
Project Management of Innovation

The selection of innovation projects is of great importance to a company that wants to stay at the front of the race with its competitors. Hence the need for innovation to be clearly defined in the company's strategy and its role in meeting the ambitions of the business to be understood at all levels in the organisation. At the strategic level there needs to be a formal mechanism for identifying any gaps in the current innovation requirements, required to meet the growth targets of the business. Once identified this innovation gap will need to be filled by the successful completion of appropriate projects that will make up the R&D portfolio. Clearly R&D will play a central role in integrating the idea generation innovation process with other functions in the company; a leading one in the work on the early stages of new product innovation and a supporting role in market development.

Some people assume that formal Project Management in R&D starts either when the experimental programme has been agreed or a product has been selected for development. This is wrong, the techniques of project management should be applied as early as possible, at least from the time an opportunity has been identified out in the market and preferably even before this, during the phase when opportunities for R&D are being sought. It is in these early phases when most projects are rejected and it is easy to dismiss an idea without giving it due consideration from a techno-commercial perspective.

Poor execution of the project selection procedure will inevitably lead to a waste of time and energy for all concerned, resulting in frustration for the research workers and the shattering of the hopes of the sponsors of the work. There is a need to look hard and long at what advantages a new product offers, how it can be assessed in terms of market acceptability and financial rewards and also in carrying out the fine tuning of the target requirements in the light of the information that is gleaned during the exercise.

Projects are either driven by the market or by a technology. In new product research the market nearly always dominates, you cannot sell a product nobody wants, whereas in process development and applications research, technological changes can more frequently drive R&D projects. Which ever type of project, a good management system will help companies to choose the right paths for their innovations and avoid them travelling down technical or commercial blind alleys. Many of these difficulties can be overcome by employing, a stage-gate process for project management. In this

Research and Development in the Chemical and Pharmaceutical Industry, Third Edition. Peter Bamfield
Copyright © 2006 WILEY-VCH Verlag GmbH & Co. KGaA, Weinheim
ISBN: 3-527-31775-9

process a gate is a milestone at which a formal decision is made on whether to proceed to the next stage. Whilst this process is good at delivery it is slow, being a linear process of around eight stages.

With the increased importance of getting new products into the market quickly, the time-to- market issue, attempts have been made to reduce the number of decisions points and stages in a streamlined process. Time equates to money, therefore the sooner a company is able to get a product to market the sooner it can recoup its expenditure on R&D. Guidelines for achieving this time reduction in the length of the innovation chain will be given, with particular emphasis on the role R&D Management plays in this process.

Selecting the Project Team and Manager or Leader, ensuring that the right people and other resources are available and assigned to the project team, are key managerial tasks. Once in place, a Project Manager, with the necessary skills and training, must have the freedom to plan, monitor and control the project and be empowered to deliver on the company's behalf a successful outcome.

Overview

The R&D portfolio, which forms part of the innovation strategy of the company, is designed to fill the innovation gap. Whilst improving existing products and increasing plant output is very cost efficient work for R&D, it is new product research, the discovery phase that underpins a growth strategy. Strategic research on the other hand generates new businesses or leads to a strategic change of direction. Successful management of the innovation chain from idea to marketable product is the key task. Full-scale manufacture of a new product is a time of high excitement, but careful planning and execution are crucial. The time-to-market issue is a driving force in new product development, but speeding up the delivery of new products to the market involves intelligent risk and resource management. Most problems occur at interfaces and are best resolved by using multi-functional project teams. The nature of the project will define the managerial skill requirements of the leader. Planning of a project and the allocation of resources should start only once it has been redefined in specific terms. A formal project launch is like the opening night of a production the successful completion of a project, and needs to be acknowledged by senior management. All projects produce new learning, which should become part of the company's knowledge management system.

1

The Selection and Evaluation of R&D Projects

R&D portfolios form part of the innovation strategy of the company

Before examining the way R&D projects are chosen and managed it is useful to re-mind ourselves of the central role that R&D plays in the totality of the company. The business processes are those of management and support for product innovation

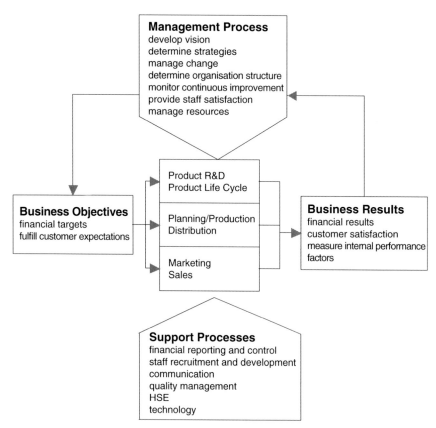

Figure D1. Business Processes

Research and Development in the Chemical and Pharmaceutical Industry, Third Edition. Peter Bamfield
Copyright © 2006 WILEY-VCH Verlag GmbH & Co. KGaA, Weinheim
ISBN: 3-527-31775-9

(R&D), production, planning, distribution, marketing and sales, which are designed to meet business objectives and produce the desired business results, as shown in Figure D1.

R&D Managers are therefore at the nerve centre of the company and must have a good understanding of the overall process for selecting and evaluating the projects for inclusion in their portfolio, and how these are developed to fit in with the company strategy. This R&D portfolio forms part of the overall innovation strategy of the company, the objective of which is defined within the strategy for the business.

1.1
Business Strategy and the Innovation Portfolio

The innovation portfolio is designed to fill the innovation gap.

There are three important elements involved in understanding the role of a business and its objectives.

1. *Having a Mission.* Often expressed in a Mission Statement.
2. *Having a Vision.* Knowing where the business wants to be in the future.
3. *Having a Strategy.* How to get from where we are now to where we want to be; achieving the Vision.

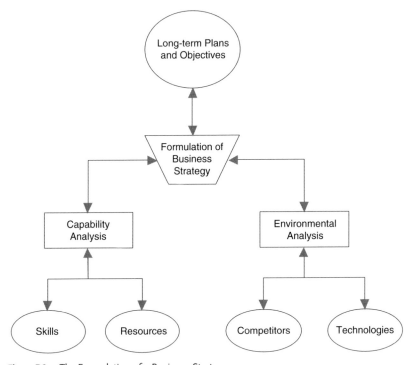

Figure D2. The Formulation of a Business Strategy

Without these three elements it is very difficult for other parts of the company, e.g. R&D, to know where they fit in and what should be their priorities.

By creating a long-term plan and setting objectives the company can formulate a strategy for the business. In order to do this it needs to look at the environment in which it is operating; the strategies of competitors, changing market conditions, technological developments, defining both threats and opportunities (see Section C, 1.6.2). It also needs to look at its current capabilities; its skills heritage and competencies; its resource capabilities, both people and financial [D-1]. The process is illustrated in Figure D2.

In the early 1990's, Kaplan and Norton devised a new strategic management tool, the *balanced scorecard*, designed to help companies clarify their vision and strategy and translate them into action [D-2]. This model contains some of the key concepts of TQM as it involves customer-defined quality, continuous improvement, employee empowerment and measurement-based management and feedback. The balanced scorecard involves developing metrics that are used to collect information on strategies and processes as data, which are analysed and fed back into the system, thus helping managers to make more effective long-term plans. This all-important feedback allows for continuous improvement of the strategic performance of a company against its vision. The methodology involves looking at the organisation from four perspectives; the internal business, financial, learning and growth, and the customer. R&D has a big input to the learning and growth perspective of this model and the sort of data that R&D might collect will be discussed later in this Section. The process is illustrated in Figure D3.

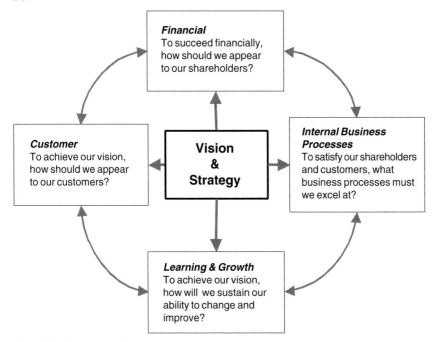

Figure D3. The Balanced Scorecard

From the financial model it is possible to identify two financial measures that are very important when looking at the extent of the need for innovation within a company. The first is the "do-nothing" model; what would be the financial consequences of just maintaining the ongoing business. The second is the so-called *innovation gap*, the financial gap between the ongoing business model and the growth ambitions of the business [D-3]. This innovation gap is the one that needs to be filled by creating a number of projects, leading to *innovation portfolio management*. The total financial rewards from the projects within the portfolio should fill the gap between the ongoing business and the growth model for the business, whilst making due allowances for the time frame and potential success rate. Consequently, the company's strategy will not only have business objectives but also an innovation/technology strategy that acts as an umbrella for the R&D strategies of each of its individual businesses.

This innovation strategy needs to be aligned to a portfolio of innovation projects, which are designed to fill the innovation gap, and which in turn can be translated into a portfolio of projects for R&D. Once again this is an iterative process involving feedback from the results, learning and knowledge gained during the innovation cycle. Identification of the specific innovation needs of the company is achieved via customer studies and idea creation, as described in Section C, 1.6.1. This generates short, medium and longer-term targets for R&D groups. In turn, work on these targets by R&D will provide results and learning, leading to new knowledge, which can be fed back into the system, thus helping to refine the innovation strategy and assist the management of the portfolio. The process is shown diagrammatically in Figure D4.

The need for an effective project evaluation methodology is highlighted by the fact that for projects in early development the odds of success are only 11%. This rises to

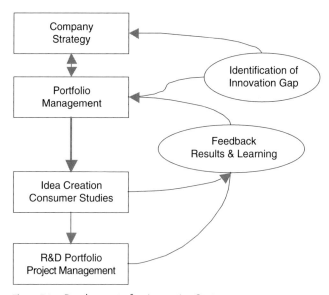

Figure D4.   Development of an Innovation Strategy

around 25% in the later development phase, as weaker projects fall by the wayside, but even after commercial launch the likelihood of success only rises to 60%. This means that 40% of projects carried to completion are a waste of money, largely due to inadequate market analysis and poor risk assessment before the start of a project. This is covered for new product developments in Section D, 1.2.2.1. Most companies use some form of stage-gate process for the management of their projects, and to assist in improving innovation are adding a discovery stage at the front end of the process (see 2.1 below).

In summary, it is essential that the following four actions are performed in developing an adequate innovation portfolio.

- Make sure that a detailed study of the market is carried out
- Perform a thorough financial and business analysis
- Carry out appropriate test marketing or field trials
- Put a maximum effort into product launch

1.2
The R&D Portfolio

The projects within the R&D portfolio, and hence the targets to be worked upon by R&D fall into the following three main categories:

1. The further development of existing products and processes in terms of cost, quality and environmental impact.
2. The invention of new products and processes for existing businesses.
3. Strategic Research, i.e. new business and technology or competency building to support or underpin the longer-term business goals.

The process that is used to generate targets for R&D personnel will depend on which of these three categories is being considered. It will also depend on the environment in which the R&D work will be carried out. Will it be within a separate, functional R&D department, as part of project team in a matrix organisation or as a fully integrated part of a strategic business unit? Whilst these latter differences are important, they are really only refinements of the general method. There are generic guidelines, which can be applied in all situations, and it is these that will be discussed in this Section.

1.2.1
The Development of Existing Products and Processes

Improving existing products and increasing plant output is very cost efficient work for R&D.

The further development of those products which are currently being sold and the processes by which they are being made, is an essential activity in maintaining the product line and the financial health of the company, whilst new products and processes are in the pipeline.

The main component of this type of developmental work is a detailed evaluation of the manufacturing processes used to make the company's products. The targets are therefore derived from discussions within the company after considering the economic climate in which the derived products and business must perform and, importantly, how customers view the current product line or ranges. The benefits that are being sought from process development work are:

- *Cost.* Reductions in the manufacturing costs and hence an improved profitability or competitiveness of the products.
- *Product Quality.* The production of consistent or improved quality products from the manufacturing processes.
- *Output.* Increased output from an existing plant or during the establishment of a new plant.
- *Environment.* A reduction in the environmental impact from the manufacturing processes.

These four areas are not independent of each other. Cost benefits can arise from producing material to consistent quality, resulting in a more effective utilisation of materials, known as *materials efficiency*, which in turn confers a reduction in the environmental impact of the process. Increasing the output, especially from an existing older plant, is very cost efficient work for R&D.

1.2.1.1 Cost Benefit Targets

The generation of targets aimed at cost reduction is an economics driven process. The targets will most likely be specific to the company's operation, since no two manufacturing plants are exactly the same, even within the same country and economic constraints. The development work will also be based on a consideration of the company's own internal information. This is because process secrecy and confidentiality means that there will be little, if any information in the external literature about a competitors process, other than the broadest outlines given in a typical process patent. It is unlikely that R&D will be able to make a detailed analysis of the strengths and weaknesses of a competitor's processes in the same way that they can of their products, samples of which are available commercially.

Following discussions with the relevant business managers, the R&D Manager is able to generate a priority list of those products that would generate most income from a reduction in the costs of their manufacture. The most basic calculation is a *cost benefit analysis.* This is an analysis of the potential financial benefits from the process development compared to the cost for carrying out the work. From these figures it is possible to determine the pay back period for the R&D investment and hence make a decision on the overall value of the proposed work. In the case of a single stage processes this calculation is relatively simple and accurate but with a multi-stage process it is much more complex.

The Manufacturing Process for FCP1
To show how this complexity arises, let us consider the manufacture of a fictional, fine chemical product used in the pharmaceutical industry, called FCP1, by a multi-stage, batch process at a rate of 100 tonnes per year.

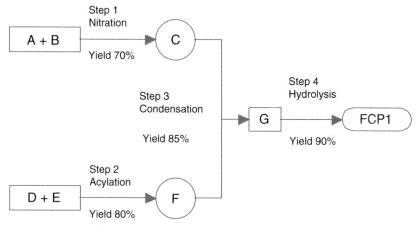

Figure D5. The Batch Process for FCP1

The route to FCP1 involves convergent pathways from the purchased raw materials A, B, D and E. In Step 1, A is used to nitrate B and produce C in 70% yield. In Step 2, Amine E is acylated with D to produce F in 80% yield. In Step 3, convergence occurs with the condensation of intermediates C and F to give G in 85% yield. Finally, in Step 4 the precursor G is hydrolysed to give FCP1 in 90% yield. The process is illustrated in Figure D5.

The cost of the raw materials are:

A £2 per kg, B £8 per kg, D £4 per kg and E £16 per kg.

The raw materials usage is as follows:
0.9kg A and 0.8kg B produce 1kg of C,
0.85kg D and 0.9kg E produce 1kg of F.
0.75kg C and 0.55kg F produce 1kg of G.
1.30kg G produces 1kg of FCP1.

Using these two sets of figures the *materials cost* for each intermediate, C, F, G and final product FCP1 can be derived:

Intermediate C = $0.9 \times 2 + 0.8 \times 8 = £8.2$ per kg
Intermediate F = $0.85 \times 4 + 0.9 \times 16 = £17.8$ per kg
Intermediate G = $0.75 \times 8.2 + 0.55 \times 17.8 = £15.94$ per kg
FCP1 = $1.3 \times 15.94 = £20.72$ per kg

Using these numbers it is possible to calculate the effect on the final cost of FCP1 by increasing the yields at one or more stages.

For instance, increasing the yield of Step 2 from 80 to 90% reduces the materials cost of F to £15.82per kg, in turn reducing G to £14.85 per kg and FCP1 to £19.246 per kg, a saving of £1.48 per kg. On a 100 tonne per year output of FCP1 this represents an annual saving of £148 000 in the materials cost of this product. If the product is sold at the same price this amount goes straight onto the bottom line profit of the business.

In addition to the savings in the cost of the final product, increasing the yield in a manufacturing process bring savings in the volumes of materials which need to be purchased.

The savings in the raw materials required to make 100 tonnes of FCP1 by a 10% increase in yield at Step 2 are:

Acylating agent D, $0.85 \times 0.55 \times 1.3 \times 100 = 61$ tonnes
Amine D, $0.9 \times 0.55 \times 1.3 = 64$ tonnes

Therefore, increasing the yield of Step 1 from 80 to 90 % requires 7 tonnes less of D, at a value of £28 000 and 7.5 tonnes less of Amine E having a value of £120 000. Total savings on materials efficiency per year is £148 000, i.e. the business does not need to spend this amount of money with the outside supplier.

The basic costs incurred for making a product, at the manufacturing plant level excluding site and company overheads, are the cost of the materials and the expenses incurred in carrying out the manufacture, including environmental charges for waste treatment or disposal.

materials cost + plant expenses = production cost

In Step 2 of FCP1, the production of Intermediate F is 72 tonnes per annum, the plant expenses being £7.5 per kg. In producing 10% more material per batch there is the opportunity to either make more of the same material from the plant or to use the spare capacity for other products. If the latter is the case, the plant expenses drop to £6.75 kg representing a saving on Intermediate F of £0.75 per kg, giving a total of *£54 000* per annum.

For a 10% improvement in yield of Intermediate F in Step 2, in manufacturing 100 tonnes per annum of FCP1, provided the product is sold at the same price and the plant capacity that is released can be filled, the total savings are:

Materials Cost to FCP1 = £150 000
Raw Materials purchases = £148 000
Plant Expenses = £54 000
Total = £352 000 per annum

The next step in the cost benefit analysis is to look in detail at the requirements for R&D effort in order to achieve the desired improvements. How many people will be required, for how long and what is the probability of success for the work.

Resource analysis for the development of FCP1 suggested that *£180 000* per annum would need to be expended on R&D, for a two year time period, to achieve a 10% yield at Step 2. The R&D spend would therefore total *£360 000*. The pay back period for a successful project would be only twelve months. The probability of success was thought to be good for a 10% and very good for a 5% improvement in the yield. On this basis the fall back position for the project would pay back the R&D expenditure in just two years. A decision was made to go ahead with the project.

Although the example of FCP1 is fictitious, it is typical of the problems associated with the definition of the targets in process development work. The complexity and inter-related nature of the many factors involved, when there are many alternative pathways, makes it a subject eminently suited to treatment by computer manipulation and commercial software is available.

1.2.1.2 Product Quality

The production of material of a consistent quality is one of the major goals of development work. Quality problems in a product are identified by the constant monitoring and analysis of the output from the plant, using statistical process control techniques [D-4]. Some of these methods have already been mentioned in Section B, 3.4.2. The avoidance of product quality problems results in direct cost benefits and also brings about a reduction in the environmental impact of its manufacture. This is because material does not need to be reworked, recycled or sent for disposal. A reduction in the number of inferior quality batches of material leads to an increase in output from the plant. More material is produced for the same effort, with the added benefit that it can be consistently supplied to the sales warehouse or be used in consuming processes.

The resolution of problems associated with the production of inferior quality products are of such importance that they must take priority over all other, non-emergency projects. Once a quality problem arises it must be dealt with at the earliest opportunity otherwise sales and customer confidence will be jeopardised. This is a special kind of development work, which is often described as *manufacturing support*. On major manufacturing plants, a local plant development team, devoted to resolving quality and production problems on that plant will carry out this support work. In smaller companies, other R&D projects will need to be delayed whilst effort is redirected to resolve an urgent quality problem.

1.2.1.3 Plant Output

Using a chemical plant to its highest efficiency and occupancy is common-sense economics. The role of R&D in this activity is to produce processes that are highly efficient, both in the utilisation of the chemical materials involved in the process and in the capital items that constitute the manufacturing plant.

The identification of the fall off in plant output uses the same statistical process control methods as for product quality [D-4]. Usually, and certainly in the larger manufacturing units, these issues will be handled by the local plant support teams. However, sometimes output issues arise which are outside the more routine evolutionary techniques employed by the process control teams. A typical example is when the output from a process is constrained by a particular plant item. An improved piece of equipment needs to be identified and evaluated. The introduction of this equipment will usually necessitate process changes for maximum efficiency. This and similar packages of work are best done by an R&D project team.

When the sales of a product are projected to reach levels beyond the capacity of the existing plant, consideration must be given to the construction of a larger one. This is an opportunity for R&D to carry out a thorough re-examination of the total process and the necessary plant, prior to the design and construction of the new plant. This is the classic case for a project team and the use of project management techniques, which will be described later in this Section.

1.2.1.4   **The Environmental Challenge**

For many years the need to have environmentally friendly products and processes was seen by many chemical companies as a necessary, but additional, cost to be borne by the business. Whilst there is still a cost, it is now more often seen as beneficial to the business and indeed to offer financial opportunities, if the challenge is accepted in an innovative fashion.

> An early example of this type of beneficial research was described by ICI [D-5]. Tioxide, then one of the subsidiaries of ICI but now part of the Huntsman Group, which makes titanium dioxide pigments, invested some £200 million over four years to reduce the environmental impact of its manufacture. One of the aims was to turn wastes into saleable products.
>
> Liquid acidic wastes from the process were neutralised to form solid gypsum that was then sold to the building, paper and farming industries. The gaseous effluent, carbon dioxide, was sold to the beer and beverage industries. Finally, the iron salts were isolated and sold to the water treatment industry. In 1994 some 580000 tonnes of these products were sold to these industries.

The environmental issues are providing an exciting opportunity for R&D to make a significant contribution, not only to the health of the local community but also the financial health of the company. Some of the major areas for study by R&D are catalytic solutions, solvent replacement, novel reactors, such as microreactors, as well as considering the product design carrying out life cycle assessments and the use of alternative feedstocks. This area has become known as Green Chemistry [D-6].

1.2.2
The Invention of New Products and Processes for Existing Businesses

New product research is the core of R&D, underpinning a company's growth strategy.

All successful new products have a period of rapid growth at some time after their introduction to the market. This is followed by a period of slower growth until the sales and profitability reach a plateau as the speciality starts to become more of a commodity product. Once it becomes a commodity that few customers want, there is an inexorable decline in both sales and profitability. The best one can hope for is that the plateau is large and that the decline into low profitability is gradual, thus delaying its eventual demise (see also Section D, 2.3.2). A product's lifetime can be extended by developing new application processes for existing areas or for exploitation in a new and growing area. Therefore, in order for it to survive in the market place a business must constantly renew its product line. Consequently, the most common research carried out within chemicals companies is new product or applications research in support of an existing business; the provision of the business with options for exploitation in the market. In pharmaceutical companies new products will be driven by developments in a particular therapeutic area, or the arrival of new diseases

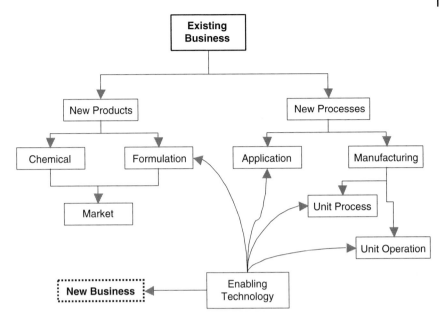

Figure D6. New Product and Process Targets

such as HIV AIDS or bacterial resistance, or social developments leading to life style aids such as sexual potency, as exemplified by Viagra and related products.

The products coming out of new product R&D work and hence its targets are almost always driven by the market and only rarely by technology. New products arising from a new external technology is generally the province of new business development or strategic research. The influence of a new, enabling technology on R&D targets for an existing business is more often found on the process side, whether it be a new application method for existing products or a new manufacturing technology. Figure D6 illustrates this interrelationship.

1.2.2.1 New Product Targets

To the R&D Manager, who is responsible for new product research, the derivation of new targets and hence research projects is an extremely important activity, as it is these targets that form the basis for all the work carried out within the R&D Group. It is only sensible that the Manager of such a group should have close familiarity with the process used by the company to define these new targets, and should be involved at all stages in the data collection and evaluation stages, and not leave these solely to the marketing and business functions.

Since the process for new products is driven by the market a business must be up to date in its understanding of the needs of both current and potential customers, since it is not a static place. All the time developments are taking place and new products are being introduced by the competition, usually without any prior warning; customers come and go and their requirements change. Information coming in from any source must be converted to knowledge (see Section C, 3.5).

There is a need to have good market intelligence, to know what is happening out there. Data on competitor's activities is more easily generated in some industry sectors than in others. For instance, in the pharmaceutical industry, the long lead time needed to meet the requirements of the regulatory authorities before the introduction of a new product, means that it is very unlikely that a new drug will be launched without the rest of the world having known about it for some time. Conversely, in the speciality chemicals or performance chemical industry sectors, a new product is nearly always a surprise. Here a myriad of new products are added each year, many of which are basically new formulations of existing products, but a significant number are new chemical entities. In these instances it is much more difficult to keep on top of what the competition is doing at any specific moment let alone what they are likely to do in the future.

In spite of these difficulties an intimate knowledge of the structures, composition and the performance of a competitor's existing products should be maintained. To do this properly requires a considerable effort particularly from R&D. A significant amount of time and money will need to be expended on the analysis, identification and evaluation of competitive products. Creating a database on the innovative activities of the competition is also useful as a benchmark for a company's own performance. However, it is mainly used as help in identifying gaps in the company's product range, which need to be filled or weaknesses removed.

The collection, evaluation and recording of data and information from the market place must be done systematically, even though a lot of it may well have been obtained in a random manner. This information will be gathered typically from the following sources.

- *Customer visits.* Especially by technical service personnel, but increasingly by directly involving R&D staff with the customer. The danger with this information is that the customer is often only interested in the immediate problems not with what can be supplied some years down the line. The data needs to be sifted intelligently to sort out the local difficulties from what are inherent problems, amenable to R&D work rather than standard technical support. A prerequisite, for all personnel who make customer visits, is the need to produce well written visit reports, with the additional requirement to report on new product options.
- *Technical symposia.* These can be very useful, more often for the "talk in the bar" rather than the technical papers, that are often very bland. Some caution is required when handling this data since it can be inaccurately reported or just hearsay and it is advisable to obtain corroborating evidence.
- *Trade Literature.* Careful reading of the trade literature can often generate ideas for new or improved products. Statistics concerning trends in technology or applications are useful when compiling the market sizes and growth potential for research targets.
- *Patent Literature.* The continuous study of the patent literature is a most important activity. It should be made the responsibility of one or more of the R&D staff who have an intimate knowledge of a particular area. The availability of computer generated profiles, down loading of data from on line access via the internet and CD-ROM archives make this a much easier task than in former years. If an informa-

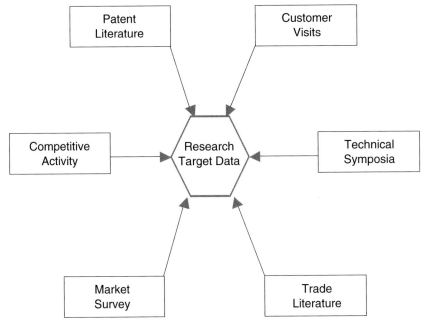

Figure D7. Information Sources for New Product Research Targets

tion scientist is available then an even more intelligent analysis can be made of the patent literature using "relational data analysis techniques" [D-7]. There is however a simple rule of thumb which can be applied by researchers. If a competitor is active in a specific area and couples this with a wide patent filing it usually indicates the nature of its next new products. Discovering a target by this method does probably mean that the required R&D is at least 18 months behind the competitor, particularly if the company is not already working in the area.

- *Market Surveys.* Such surveys carried out by consultants are only of value in confirming general trends. They are very unlikely to identify the need for a specific product. It is much better for a company to produce its own market survey having gone to the trouble of generating the data from the above sources.

The inputs for generating research targets, summarised in Figure D7, are both explicit and tacit, as explained under knowledge management (Section C, 3.5.1).

Having generated and analysed this information it is possible to draw conclusions about the potential targets for the R&D group. A full evaluation of these potential targets is then carried out before any are converted to projects. This evaluation will involve both economical and technical issues. The questions that will need to be answered are as follows.

Economic
- *Market size.* What is the total market size in the area the product will be sold, how fast is it growing?

- *Market share.* Forecast percentage share for new product; what share of this existing market is the product likely to take?
- *Sales forecast and growth.* What volume will be sold and what will be the rate of penetration of the product in the market?
- *Selling price.* What is the likely selling price in the range of markets in which it will be sold?
- *Profitability.* What profitability is required?
- *Target manufacturing cost.* What is the maximum that can be afforded in order to meet the profitability criteria?
- *R&D costs.* How many people will be required on the project and how long will it take?
- *Launch costs.* What will be the costs of the initial manufacture and the launch onto the market?
- *Capital requirements.* Will the profitability of the product justify capital investment if required?
- *Return on investment.* How quickly will there be a return on the investment, what will be the pay back period?

Technical
- *Performance criteria.* What are the performance criteria that the product must meet, which of these are absolutely essential and which are desirable to have?
- *Competitive Products.* With which products will the new product have to compete, what are their strengths and weaknesses?
- *Likelihood of success.* What is the estimate of the research producing a product that will meet all the economic and performance criteria?
- *Technological compatibility.* Is the product likely to be one that can be made in existing plant or by current technology?

These are generic questions, which will be modified to fit the requirements of a particular market or business area. For instance, a research proposal for a new drug would have to consider its performance in a specific therapeutic area and whether it genuinely will meet the unmet needs of patients. Regulatory authorities are demanding that a new drug show significant advantages over existing drugs, hence the performance level of any new entity must be agreed before any work is begun.

After evaluation of the potential targets against these criteria a list of R&D projects can then be drawn up in collaboration with the marketing function of the business, creating a project portfolio. On the basis of this detailed assessment, each of these projects will have listed the economic and performance criteria that they must meet to be successful. It is important that all those concerned with making a decision on the outcome should agree that these criteria are correct, and that they are not going to be changed significantly during the research phase. It is very difficult to plan target-orientated research if the goal posts are always being moved because they were not in the right place in the first instance.

The number of people allocated to each project, i.e. the proportion of the research budget, will depend largely on the size of the eventual economic return set against a desired pay back time. If getting there quickly is of overriding importance then more people will be required, provided this fits in with the perceived ease of achieving the desired target. This is considered in Section D, 2.3.2.

1.2.2.2 New Process Targets

There are two main types of process research undertaken in order to grow an existing business. The first is designed to expand the manufacturing capability of the organisation and the second is that which looks at new methods of utilising or applying existing products in the hands of the customers.

New Manufacturing Processes

Research into new manufacturing processes is most commonly carried out in multinationals or those companies whose main or sole business is in contract manufacturing area. These companies operate in both the fine chemical and the bulk manufacturing sectors. New process research in these companies has the following driving forces:

- *Capacity.* To fill the capacity of a plant with products, the sales of which are constrained by material availability, but whose processes cannot be directly transferred to another plant.
- *New products.* To produce for sale products or intermediates which have a good market but which are currently not on the company's selling range.
- *New technology.* To take advantage of a technological development, in the form of a new unit process or operation, which could be used to make new products for the market place.

The identification and evaluation procedure for these research targets is essentially the same as is used for new products. New technology is increasingly being developed in smaller companies with a specific competence. An example would be the work on production scale microreactors by companies such as CPC Cellular Process Chemistry Systems, whose technology is being taken up by large companies in the pilot scale manufacture of fine, speciality and medicinal chemicals.

New Application Processes

A very cost effective way of expanding the sales of a product or product range is to find new ways of applying these materials in the hands of the customer. A new application method can take the form of an improvement over an existing one, giving the company an edge over the competition, or it can be for an entirely new application and hence new business. Companies in the whole of the speciality and performance chemical areas lean heavily on novel application type of R&D work to expand their business. It is cost effective because it does not require the invention of a new chemical and the attendant costs of process development and registration. In the pharmaceutical area, a new method of delivering a drug is an example where maximum benefit can be obtained from an already approved chemical entity. It is uncommon for a product devised for one application to be directly of use in a new application process. Some modification is usually required. This often involves either reformulating or purifying the product to meet the different requirements of the new outlet. An example of the latter case was the use of highly purified textile type dyes or pigments as colorants in inks for the early inkjet printers for home-office use.

1.3
Strategic Research Targets

Strategic research either generates new businesses or leads to a strategic change of direction.

R&D plans are constructed so that they fit in with the overall innovation strategy of the business, which will have a component covering technology strategy as well as a new product strategy. The business strategies will contain short, medium and long-term goals. The R&D work described in Section D, 1.2.1 and 1.2.2 is carried out to support the short and medium requirements of the business. The longer-term aspects are generally part of the technology strategy, and form the basis of what is often described by the term "strategic research", which can be classified into two generic types:

1. Research designed to use *existing competencies* in a new way and hence to generate *new businesses*.
2. Research done in order to *build the competencies* required for a strategic thrust in a *new direction* for the company.

The relationships between the business strategy, new product research and strategic research are shown in Figure D8.

The generation of new businesses on a base of existing competencies starts with an examination of the core strengths of the company. Once these core competencies are fully understood it is possible for the strategic planners to look at their wider application. Simplistically this is done by listing those new or emerging areas of business where the company could take advantage of these competencies. The classic case in the Chemical Industry, was the move made by the larger European chemical companies, over fifty years or more ago, into medicinal and agrochemicals, based on their skills in fine chemicals built over generations in the dyestuffs and pigments industry. From these tentative beginnings extremely large companies, such as AstraZeneca and Syngenta, have emerged decades later.

The building of competencies for a strategic move into entirely new areas is often driven by developments outside the chemical industry. For instance, the 1970s saw a boom in the electronics companies based on silicon chip technology and all the related areas which flowed from this development. This encouraged many companies to move into the chemicals for electronics areas. Thus there was a need to build competencies in solid-state chemistry and physics, materials science, clean room technology etc. In the 1980s the emerging technology was biotechnology and this also required new competencies, such as molecular biology. In the early 1990s pharmaceutical companies were building the skills required for combinatorial chemistry to meet the needs of high throughput screening technologies. Similar competency building is occurring now in the developing areas of bioengineering, genomics, nanotechnology and new energy sources, such as fuel cells.

Many large companies find it difficult to manage these competency-building activities, and many have closed down their corporate laboratories, as discussed in Section B, 1.1.1.1. Hence, these competencies are often developed in smaller, start up companies sponsored by venture capitalists. These start up businesses then license

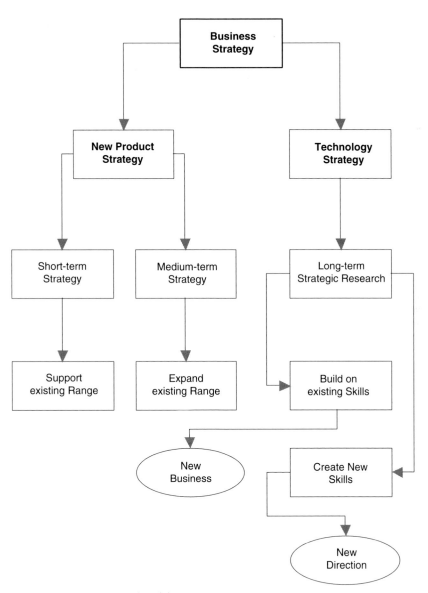

Figure D8. Strategic Research and the Business Strategy

or transfer their technology to the larger companies or are in turn acquired so that the competency is then achieved by acquisition (see Section C, 3.4). An alternative approach is to outsource this type of work to research consultancy companies, universities and institutes. This is especially true when the work requires some fundamental understanding to be obtained before any significant progress can be made. These collaborative contracts are a special feature of the outsourcing of R&D (see Section B, 1.4).

Strategic research works on a five to ten year minimum time frame and any business option must be judged on what it will be like in ten years time. By its futuristic nature the outcome from strategic research is difficult to evaluate commercially. The error bar, especially in the emerging business area, is considerable and hence the need for this work to be sponsored outside the normal business route for R&D. It has to be protected from the vagaries of the business cycle and must have a strong sponsor on any Executive for it to survive.

The day-to-day management of strategic R&D requires the deployment of special managerial skills. Patience is always required, the long view must be taken and the researchers encouraged at all times, in what is a very exposed position for them to work. The Manager must be on the look out for the unexpected to happen and when it does be ready to exploit any reasonable opportunity. The exploitation of the results of strategic research requires an entrepreneurial skill not often associated in people's minds with those carrying out long-term research. It is therefore quite common to have another person, having the creative skills to carry out this role, identified as a *new business manager*.

2
The Innovation Chain

Successful management of the innovation chain from idea to marketable product in the shortest possible time is the key task.

The *Innovation Chain* begins with the identification of an opportunity for the business and ends with a product or range of products available for sale. Whilst the definition of the opportunity, the construction of the R&D target, is a key action, it is only the first in a series that are required in the process of innovation. The task for the company is to manage the overall process through to a successful conclusion in as short a possible time. This is the arena in which project management operates. It is not the purpose of this book to try and define the best formal project management system. Most companies will have one that they prefer to use in any case. The objective in this Section is to help Managers to get an overall view and understanding of the total process, how it applies to the Chemical and Pharmaceutical Industries, and to offer advice on how to avoid the many pitfalls for the unwary. The British Standard Institution (BSI) has issued a guide to project management BS6079, which describes a full range of project management procedures, techniques and tools, and which is worth exploring by those requiring greater detail [D-18].

2.1
The Stage-Gate Process

As mentioned earlier the odds of failure for a new product are high. For every eleven new products in the pipeline only three enter the development phase, only 1.3 are launched and only one is successful. One way of improving the chances of those entering into development getting through to launch and then becoming successful products is to use an effective project management methodology. The majority of companies use some form of stage-gate process, the most common of which was devised by Bob Cooper in 1993, who later incorporated project selection and portfolio management into the version published in 2001 [D-8].

In a stage gate process, gates are decision points where the management team reviews progress and decides whether sufficient has been achieved and if sufficient information is available to sanction progress to the next stage. If things are not looking good they can request that more work be done prior to making a decision or they may

Research and Development in the Chemical and Pharmaceutical Industry, Third Edition. Peter Bamfield
Copyright © 2006 WILEY-VCH Verlag GmbH & Co. KGaA, Weinheim
ISBN: 3-527-31775-9

decide to cut their losses and abandon a project. The decision to abandon should not be seen as something to be ashamed of, but indeed it should be looked at in a positive light as money saved. The stage-gate process is illustrated in Figure D9.

2.2
New Chemical Product Innovation Chain

The various steps in the chain leading to the introduction of a new chemical entity to the markets of an existing business are shown in outline in Figure D10. These are the basis for a stage-gate development plan at the individual chemical product level. The basic steps following the target definition are:

- Speculative research for the new chemical
- Development of the process for its manufacture
- The first manufacture of the product
- The launch of the new product into the market
- The further establishment of the product and process

There are many crucial activities which must be completed during or in parallel to each of these phases, such as application testing, HSE studies, product registration and customer, clinical or field trials, these are discussed in greater detail in subsequent sections. If all the steps were carried out in a sequential manner, the time from start to finish would be inordinately long. This is never the case, most of the steps overlap and the big issue of the management of time in the overall project is covered in Section D, 2.3.

2.2.1
Speculative Research

The details of the new product, which is the target for the speculative research chemist, will have been defined in the research target document, using the methodology described in Section D, 1.2.2.1. Provided this has been done correctly, the people making the new chemicals and those whose job it is to evaluate these materials will be working to the same criteria. This may seem obvious but it is the obvious that is often overlooked. Managers must never let researchers start work on a project until they are satisfied with every aspect of the target document.

Initially, the criteria used to evaluate the samples produced in the laboratory will be those based on the desired technical performance. However, if the performance criteria for a new product have been defined in their entirety, it is unlikely that any speculative sample will satisfy all of them. Therefore, it is usual to define a default set of essential properties in the target document. A problem arises if the best candidate products meet all of these but misses on one or more of the important but less essential other criteria. In this case a view needs to be taken on the relative importance of the criteria that have not been achieved, and a decision made on which product, on balance, is the best. The defining question is "will this product be good enough

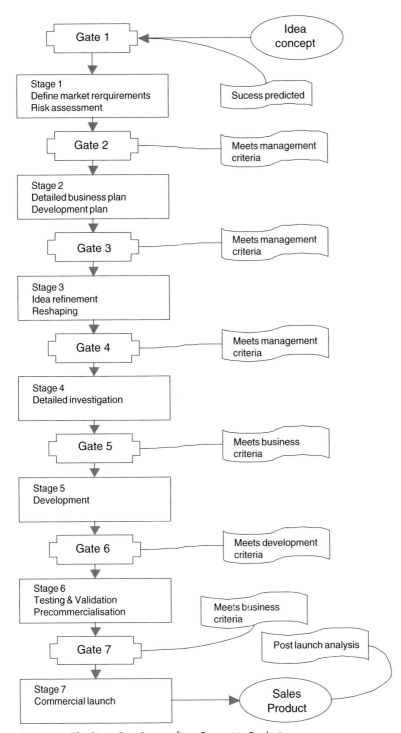

Figure D9. The Stage-Gate Process from Concept to Product

to command a presence in the market?" Or alternatively "does the product show sufficient advantages over the competition to justify development?" There is little point in aiming for perfection if the additional sales generated will only be marginally beneficial, and if the research phase will become protracted and a market launch delayed.

If the answer to the question on the market potential of a candidate is positive, a preliminary view of the cost of its manufacture is required. This figure will in turn

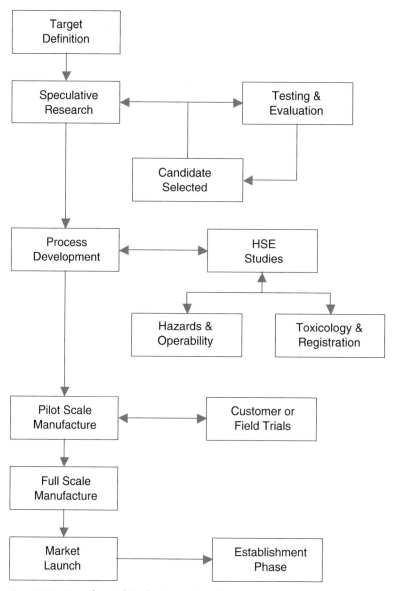

Figure D10. New Chemical Product Innovation Chain

enable an estimate of the potential profitability of the product to be made, and compared to the financial criteria set out in the target document. Process details are likely to be sparse at this stage and an analogous, existing process will be used as a basis for the calculation. If the estimated costs and profitability are marginal to the requirements, or if a good analogy does not exist, it is better to suspend judgement until some process development work has been done in the laboratory. It is also wise at this early stage, to seek an expert view on whether there are any potential toxicological problems with the chemical or any of the intermediates in the manufacturing route. Any doubts in this area justify expenditure on preliminary toxicity screening. This prevents the waste of research resources that can happen if a toxicity problem is discovered later in the development cycle, during full testing, leading to abandonment of the project. Similarly, a view should be sought on any safety or plant operational hazards in the projected manufacturing process.

When the prototype product satisfies all the requirements for the addition of a new product to the range, set by the business, it is ready for further development.

2.2.2
Process Development

The development of a process to make a new chemical entity is not a simple procedure and a project leader will need to be appointed, if one has not been there during the speculative research phase. Process development usually costs several times more than the expenditure on the initial product research and is not entered into lightly. The procedure adopted for process development needs thoughtful planning, if time and money are not to be wasted. The procedure, which is described here, is normally applied to novel chemical entities, but many of the aspects apply equally to the addition of a known material to a company's selling range.

Optimisation of the materials usage, yield, time and other parameters of the laboratory method, used by the research chemist to make the original speculative sample, will not have been carried out at this stage in the project. Additionally the process, even if a good one for the laboratory, will not necessarily be one which will work or fit into an existing manufacturing plant. The first task of the project leader along with the development chemist or Manager is to ask a series of questions on the characteristics of the process that has so far been used to make the new product.

- *Chemical unit processes.* Are they ones that are currently operated on the plant or are there good analogies? For example, if it involves the current nitration, reduction, chlorination and enzymatic catalytic oxidation processes operated in the company, the available chemical experience will be satisfactory for the development. If, however, it involves highly reactive organometallic reagents, beyond the current expertise of the company, new learning will be required or experience brought into the team from elsewhere or that stage outsourced to another group or company.
- *Unit operations.* Are the unit operations ones that are currently used, e.g. distillation, fermentation and freeze-drying? If it is new to the plant, is the necessary chemical engineering expertise to implement the operation available?

- *Plant requirements.* Is it a single stage, multi stage batch, continuous process or any combination of these? Will it use the existing plant or is a new plant required? If it is the latter, chemical engineers and a design team will be involved.
- *Manufacture.* Will the first manufacture be on a commercial scale or will there be a need for pilot plant quantities of material.

The answers to the above questions will enable the project leader, in collaboration with the relevant Managers, to assign the correct personnel to the project at the start of the development work and to plan for additions later on as required.

The performance criteria on product quality for the development candidate may have changed marginally since the production of the original target document. Even if they have not changed, it is good to reiterate the performance criteria so that they are fully understood by all the new parties who are involved. The details in any statement of profitability criteria, especially the desired manufacturing cost, will have become firmer by this stage and agreed with the business function. An early indication of whether capital will be available for any new plant or plant items is also necessary, as its availability will materially affect the decisions made during the process development phase.

2.2.2.1 HSE and Product Registration

The health, safety and environmental (HSE) aspects of the chemical manufacture should be uppermost in the team's mind throughout the course of the development. However, once the development of the process has reached the stage when only very minor modifications are likely to occur, it is time to look at the formal aspects of HSE. The components of HSE that are internal to the company are:

- Chemical reaction hazards
- Plant operational hazards
- Health hazards in the workplace
- Environmental impact of manufacture

The evaluation of the potential hazards associated with chemical processes and plant operational procedures is a specialised activity. In larger companies a separate group will be responsible for this activity. The work involves testing for any fire and explosion hazards associated with the chemicals and the reaction mass in which they are produced, known as the *chemical reaction hazards*. The process details are considered in relation to the actual chemical plant in which they are to be operated; the *plant operational hazards*.

Estimating the environmental impact of a manufacturing process involves its assessment in terms of all the possible emissions, whether they are gaseous, liquid or solid. One of the objectives during process development work is to eliminate any potentially noxious emissions and to minimise all others. If it is not possible to minimise, say liquid or solid wastes, they should be rendered benign before disposal. In most countries the polluter must, quite rightly, meet the cost of any pollution control. There is a clear economic argument for companies to use clean processes as well as moral and legislative imperatives against causing any pollution.

The control of health hazards in the workplace is subject to legislation as outlined in Section B, 3.2. This involves an evaluation, by an occupational health expert, of the toxicity data available on all chemicals, including purchased intermediates and reagents. The final product will of course need to be tested for registration purposes and this is outlined below. Data will also need to be generated on all other intermediates for which there are no known toxicity information. On the basis of this information exposure limits can be set for workers and the need for protection measures during processing defined. Such work can, and indeed often does, lead to a need to change the process in order to avoid an intermediate or unit operation that is problematic, in terms of work place protection [D-9].

Before a material can be sold on the open market it must undergo *Product Registration*, or Notification as it also called, with the relevant regulatory bodies. It should be noted here that in pharmaceutical companies a massive amount of work is required to register a new drug, taking many years and involving whole departments. (See Section B, Figure B8) Preclinical development of a drug is carried out to collect sufficient information on the safety, ancillary pharmacology, disposition and side effects following administration of a new product to animals that will allow the company to progress into its testing in humans' Phase 1 of clinical trials [D-19]. It is beyond the scope of this book to deal with such aspects of registration for a new drug, and the description of the regulatory process given here is typical of that for a speciality chemical. A Regulatory Affairs manager will handle this work in most companies of any size.

In the European Union (EU) there is a list of chemicals called EINECS (the European Inventory of Existing Commercial Chemical Substances). This covers all chemicals that were on the EU market prior to 1981. Until recently, only new substances, defined as those that are not on EINECS list, were required to be registered and the necessary information supplied to the regulatory authorities. In the USA this is covered by the Toxic Substances Control Act (TOSCA), and in Japan by the Ministry of Technology and Industry (MITI). However, in 2001 the EU proposed a policy, with the acronym REACH, under which ALL chemicals, even those on the EINECS list, manufactured or supplied in the EU must eventually be registered, involving a strict and costly testing regime. The EU outlines the REACH proposal as follows.

"The REACH proposal gives greater responsibility to industry to manage the risks from chemicals and to provide safety information on the substances. Manufacturers and importers will be required to gather information on the properties of their substances, which will help them manage them safely, and to register the information in a central database. A Chemicals Agency will act as the central point in the REACH system: it will run the databases necessary to operate the system, co-ordinate the in-depth evaluation of suspicious chemicals and run a public database in which consumers and professionals can find hazard information [D-20]".

The original proposal met with considerable opposition from the chemical industry and lead to an amended proposal in 2003, political agreement was obtained in 2005 and implementation is planned for 2007. A new European Chemicals Agency is expected to be fully operation 12 months later.

The acronym REACH stands for the following:

- **R**egistration of all substances >1 tonne per year
- **E**valuation of substances >100 tonnes per year
- **A**uthorisation of **CH**emicals of special concern

The guiding principles of REACH are:

- Protection of human health and the environment
- Maintenance/enhancement of competitiveness in the European chemical industry
- To prevent fragmentation of the internal market
- Increased transparency
- Integration with internal efforts
- Promotion of non-animal testing
- Conformity with EU internal obligations
- Precaution
- Substitution

The deadlines for registration are set according to the volume of the substance on the market or the hazard. The shortest deadlines apply to very high volume substances (above 1000 tonnes), and carcinogenic, mutagenic or reproduction toxic substances above 1 tonne. These will have to be registered within 3 years.

Returning to our genuine new product, in the EU, the Notification requires a Base Set of information, which includes the following [D-21]:

- *Physical chemistry*. Identity and physico-chemical properties of the substance. This includes the structural formula of the chemical or the components of a mixture, melting point, boiling point, water and fat solubility, vapour pressure, partition coefficient and data on flammability and explosion characteristics. This involves a great deal of detailed analytical work all of which must be carried out to the highest standard, for instance using Good Laboratory Practice (GLP) in a quality assured laboratory.
- *Toxicology*. Acute toxicity studies involving oral, inhalation and dermal (LD50 and LC_{50} figures), skin and eye irritation and sensitisation, sub-acute effects and screening for mutagenic and potential carcinogenic activity, both in-vitro (e.g. Ames Test) and in-vivo (e.g. mouse micronucleus test). These tests will be carried out in a laboratory specifically skilled in the task. Increasingly this work is done under contract rather than in house, except in the largest of organisations.
- *Ecotoxicology*. This includes an investigation of acute aquatic toxicity (e.g. on fish and daphnia) and biodegradation potential and biological oxygen demand (BOD). As with toxicological studies these tests need to be carried out in a recognised expert laboratory.

The tests on new materials are both extensive and expensive. For a typical speciality chemical in 2001 this amounted to around 150–200000 euros, adding anything between 14–38% to the development costs of a new product. They therefore should only be undertaken using material, which has been produced in consistent quality, in terms of chemical composition and physical form, by the process to be used in a full scale or pilot plant manufacture.

A limited announcement package is possible for small quantities, e.g. sales of less than 1 tonne per year. This allows a new product to be launched in a test market mode. The test required by the Regulatory authorities increases as the scale of manufacture increases from 10 tonnes per year to 100 tonnes and then to 1000 tonnes per year to a maximum at 5000 tonnes.

2.2.2.2 Pilot Plant Manufacture

The factors affecting the performance of chemical processes, on scaling up from the laboratory process to the full manufacturing scale, are many fold and their behaviour often difficult to predict. The intermediate step between the laboratory and the full scale is the pilot plant. The use of a pilot plant trial has two useful technical benefits. Firstly, it provides the opportunity to test out the process at a somewhat larger scale, to check process variables, and provide confidence about the performance of the process on the full-scale plant. Secondly, it allows larger samples to be made for toxicological testing and for trials in the field and with customers on a commercial scale. For the R&D Manager or Project Leader, the pilot plant trial offers the opportunity to try out and train the team that will supervise the technical aspects of the first, full scale manufacture. The Manager should be looking to see how well each person in the team relates and whether adequate technical cover is being provided at each crucial step.

2.2.2.3 Field and Customer Trials

The term *field trial* is often reserved for the specific task of evaluating agrochemicals on the farm. However, it can also be more generally applied to those trials on a larger scale with customers. The nature of the field or customer trial will determine the type of R&D personnel who are involved. For instance, in agrochemical companies, biologists and botanists usually supervise field trials, whilst the evaluation of anticorrosion chemicals on oilrigs is likely to be done by a chemical engineer. Technical service personnel will carry out trials on products directly destined for the consumer, for instance the dyeing of fabrics in a dyehouse.

Close collaboration between the Project Manager, the project team and those involved in the organising of field and customer trials is necessary. This needs to be done in order to deliver the bulk material in time to meet certain deadlines. Examples include the need to have material available to spray crops during the growing season, having samples ready to go in to a long term testing trial for corrosion protection or available to produce a new paint formulation in a customer's plant and to use this at a trade show.

The results from field and customer trials can be immediate, as in the case of the paint formulation, or take a long time to obtain, as in the evaluation of corrosion protection. Because of the time and cost involved, it is unlikely that field trials would be organised without a high probability of success. Whilst awaiting the results the time can be usefully filled with work on the final process and plant refinements for the full scale, trial manufacture.

2.2.2.4   **Clinical Development**

Far more complicated, and greatly more expensive, than customer and field trials of a speciality chemical is the clinical development work required for a new drug.

Phase I trials are the first-stage of testing in human subjects. These tests take about a year and involve about 20 to 80 normal, healthy volunteers. This phase includes trials designed to assess the safety profile, including the safe dosage range and tolerability. The studies also determine how a drug is absorbed, distributed, metabolised and excreted, and the duration of its action.

Phase II trials are controlled studies of approximately 100 to 300 volunteer patients (people with the disease) to assess the drug's effectiveness. These trials can take about two years.

Phase III studies are double-blind randomized controlled trials on large patient groups (1000-3000 or more) in clinics and hospitals. Physicians monitor patients closely to determine efficacy in comparison with currently available alternatives and identify adverse reactions. Phase III trials are the most expensive, time-consuming, they can take three years, and difficult trials to design and run.

Clearly this involves a major commitment of time and money on behalf of the company and will clearly be managed by as a project with a team and manager for the various phases.

2.2.3
Trial Manufacture

Full scale manufacture of a new product is a time of high excitement.

The planning and execution of the full-scale manufacture of a new product is a time of high excitement in R&D. It is often the culmination of several years work, involving many people from different disciplines. Everybody is keen for it to succeed and the profile of this activity within the company is very high. It is the time when senior manager's interest is aroused and this can put some pressure on the team. The R&D Manager needs to be especially sympathetic and supportive of the project leaders actions during this period.

If the manufacture is in a multi purpose plant the project will have been allocated a time slot within which to complete the work. Good planning of the allocation of personnel, their organisation into teams, usually working in 24-hour shift teams, is a key responsibility of the R&D Manager or the project leader. A rapport with the plant management team must be established as their help will be needed, no doubt at short notice, even with the best project management system in place. The analytical support must be energised and the product evaluation team encouraged giving a rapid response on product quality. The team leadership and motivational skills of the Manager need to be brought into full play during this period.

Manufacture of any pharmaceutical product must satisfy the requirements of Good Manufacturing Practice (GMP). These are basically that the equipment be properly designed, maintained and cleaned, that Standard Operating Procedures are written, approved and followed, quality to be independently monitored and that all personnel are properly trained. (See Section B, 3.3.2.3). Management must ensure that there is absolutely no slippage in the requirements of GMP.

2.2.4
Market Launch

This is the crucial test. After all the planning, designing and manufacture will the new product be accepted with enthusiasm by the market? Will it generate money that will start to pay back the investment made by the company in R&D?

The marketing department drives the market launch but they need, and should receive, the full support of R&D. This is provided in the form of technical support to the sales force and assistance to customers when they are first using the new product, anticipating and responding quickly to any unforeseen problems that might arise.

2.2.5
Establishment of the Manufacture and Sales

Once a product has been launched R&D's involvement is not over. The process for the initial manufacture needs to be improved, the learning curve followed so that costs are lowered and the profitability of the product increased. A systematic process improvement programme needs to be planned by R&D and supported by the business (see also Section D, 1.2).

Out in the market place, the product will need to be further differentiated from the competitive products or modified to meet the challenge, which will no doubt come from other companies as the new product eats into the sales of the existing ones. Typical modifications that occur in the early life of a new speciality chemical are changes in the physical form or fine tuning of the application method as experience is gained in the field. For instance, the initial material may have been launched as a free flowing solid but some customers require it to be in a liquid or a pelletised form for ease of handling. R&D will be required to produce these various physical forms to meet the customer's needs. Formulation science and technology is therefore a very important competency in those industry sectors where products are sold for their effect or performance in the hand of the customer. In order to meet the full market potential modifications to the application process may also be needed to take advantage of the expertise and skills existing in customer companies.

2.3
Time Management of the Innovation Chain

The time-to-market issue is a driving force in new product development.

The time taken to travel along the innovation chain (see Figure D10), known as the *lead-time,* can vary from several months in speciality chemical companies to many years in pharmaceutical companies. This wide variation can, in part, be explained by straightforward technical differences due to legislative impact and the time taken to evaluate products, either clinically or in the field. For instance, it only takes months to put out a new speciality chemical product, which is a mixture of known ingredients, but many years for a new medicinal chemical with all the clinical trials and the regulatory requirements that are required as already described in Section D, 2.2.2.4. In spite of the vast technical differences which exist between one sector and another, in all cases the process is capable of being shortened by an examination of the non technical issues, and the effective management of time within the specific project.

The time taken to get a product from the target definition stage to the date when a product is available for release into the market, i.e. to complete the innovation chain, is known as the *Time-to-Market*. For over a decade, the time-to-market issue has been a driving force in new product development, especially in the pharmaceutical industry, and its reduction is still very high on the list of a company's strategic objectives [D-10, D-11]. There are two main reasons for this desire to reduce the time-to-market:

1. Faster evaluation of a new product in the real world of the market.
2. Faster return on the money invested during development.

To achieve the benefits which can accrue to the company, the emphasis must be on speed; speed in R&D, through rapid development and hence production, and speed in marketing by having a rapid sales response and prompt customer service.

2.3.1
The Faster Evaluation of New Products

The lifetime of new products in the technological markets has become shorter and the turn over very high as innovation has become a main driving force. This has been especially true for in the new products generated by the electronics business and related areas; computers, telecommunications, mobile phones, flat screen TVs and peripheral technologies such as digital recording and imaging, colour printing and speciality papers and, especially for use in the office and the home. This rapid change has spilled over into many other sectors of business, particularly those with a strong fashion element. These rapid changes in product lines have therefore required equally rapid responses from those companies that are active in supplying chemicals into these consumer areas.

One of the driving forces, behind this need for the rapid introduction of new products, has been the belief that spending a lot of time on detailed market research is less accurate, and in the end more costly, than getting a product quickly out into the market and letting the consumers pass judgement. Thus, the really fast innovators show the following behavioural patterns:

- They use their customers to fine tune their innovation based on rapid feed back from the market to the designers.
- They provide less improvement with each product but introduce new products more frequently.
- They devise patterns of incremental experimentation to achieve a competitive edge.

This method or style of introducing new products, which is common with companies introducing new consumer goods, machines or devices, is very difficult to achieve in the chemical field. This is simply because of the time that is required to satisfy the legislative framework, which applies to all new chemical entities (see Section D, 2.2.2.1). However, it is a method that has been practised routinely by those chemical companies, often known as formulators, whose development work is centred on making new formulations of existing materials to meet a specific customer's specification. Rapid introduction can also be achieved with new application methods for products that are already on the selling range, and have therefore satisfied the regulatory authorities. The real need, in whichever area of the chemical or pharmaceutical business the company operates, is to innovate faster then the competition, to get new products out to the customer faster than they can, giving itself a competitive edge.

2.3.2
The Faster and Increased Return on Investment

Whilst there are external restrictions on the speed a new product can be introduced into the market, as outlined in Section D, 2.3.1, the major limitations are set by a company's own internal systems and methods of project management. A goal for every R&D Manager must be to overcome these internal limitations, and to get a faster return on the company's investment in R&D. The financial returns to the company, from a more rapid development, can be easily quantified and the figures used to justify an increase in R&D effort to attain this objective.

Any reduction in the lead-time, from the target setting stage, through the research and development and manufacturing phases to product launch, brings both intangible benefits, such as being seen by the customer as both innovative and fleet of foot, and tangible benefits for the company, for example:

- Money is coming in from the sale of the product sooner than would otherwise be the case.
- Higher prices can be obtained for a new product for a longer period before the competition arrives.
- The profitable stage in the product life cycle is extended resulting in an increased pay back.

The increased sales volume obtained from the rapid introduction and extended product life cycles is illustrated in Figure D11. The time period 0-A is the introduction phase when sales are low, keeping this a short as possible by a having well planned product launch is very important. During the period from A-B, providing the product has been designed to meet a correctly identified customer need, sales should be growing rapidly. The rapid evaluation of the product, described in Section D, 2.3.1, occurs during the period from 0 to some time during the early part of B. This is when an early decision can be made on whether to abandon or continue with the new product. The mature sales period for the product is from B-C, thereafter there is usually a steady decline in both sales volume and profitability. The area between the two graphs represents difference in sales volume between the slow and fast innovators.

The profits from a new product reach a peak before the time its sales are at a maximum, as illustrated in Figure D12. This is because, as stated above, a premium price can be obtained in the market until the point when competitive products arrive. From that point onwards prices start to drop even though the market size is growing. This is the period when there is maximum pressure to cut costs and to introduce more efficient manufacturing processes, work that should have been going on in R&D since the product was first launched. It is also the time when a new product should be in the pipeline, ready for launch before profits and sales start to decline on maturity of the existing product.

Another way of looking at the value of a more rapid introduction of new products is to examine its effect on cash flow. The cash flow in the project is increasingly negative during the research, development, make and the initial sales period, represented by A-C in Figure D13, thereafter this is reversed and the flow is positive, growing in line with the sales revenue. If the lead-time is reduced, A-B on the graph, the cash flow is negative for a shorter period and often, but not always, smaller in total.

Whilst the metrics, shown in Figures D11, D12 and D13 are very useful in calculating the possible effect of shortening the lead times, they are not of any direct value in determining the effect of expenditure during a project. A methodology has been described to estimate the effects of such expenditure, using what is called a *Return Map* [D-12]. The main use of the Return Map is in enabling the members of a project team to as-

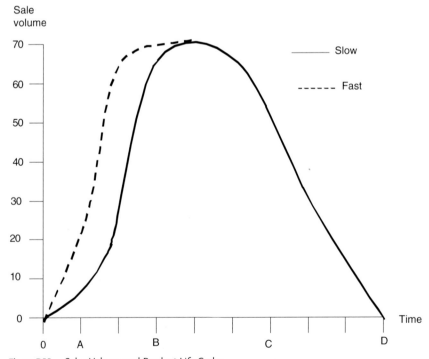

Figure D11. Sales Volume and Product Life Cycles

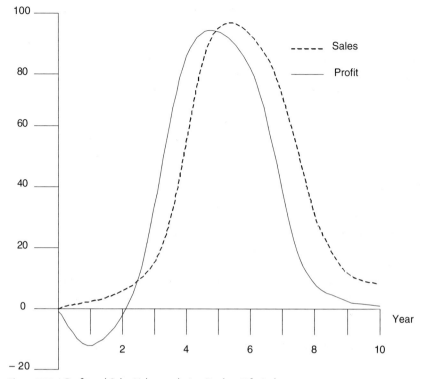

Figure D12. Profit and Sales Volumes during Product Life Cycles

sess their contributions to the success of the product in terms of both time and money. Its key difference with other metrics is that it uses as the goal the time it actually takes to break even on the project investment, rather than a simple time to market figure. A much-simplified version, applying terms already used, is shown in Figure D14.

The main element of the Return Map is a graphical representation of the estimated expenditure or investment, the sales income and the profits plotted against time. The time-to-market is the time A-B; the time to break even after product has been launched is B-C, whilst the overall time to break even is A-C. This is the basic set that is used by the team at the beginning of the project. As the project proceeds it is modified in the light of experience and the time to break even, the key metric, is further refined. It is then possible to ask "what if" questions and measure the effect on the break even time. For instance, it could emerge that by spending a significant amount of extra money during the R&D phase, the time-to-market could be reduced by six months. Using the Return Map it is possible to calculate whether the time to break even is also shortened. If it is not shortened then, on this basis, the increased investment is probably not justified.

One of the big advantages of using methods like the Return Map is that they require inputs from all the functions that will contribute to the project. Not only must R&D provide the development data, but also manufacturing must be involved in fore-

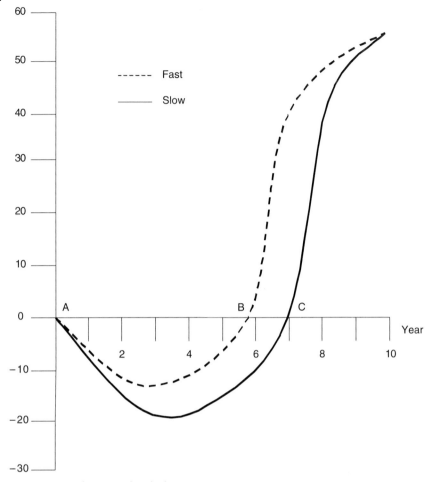

Figure D13. Lead Times and Cash Flow

casting capital and manufacturing costs and marketing are required to provide accurate sales forecasts. This involvement means that all parties think deeply about their contribution and become more committed to achieving the goals of the project.

2.3.3
Lead Times and the Development Risks

There is a correlation between the cost to the business of a delay in the lead-time for a new product and any risk to the success of this product in terms of its functional design. This is shown schematically in Figure D15.

At one extreme of this correlation, the risks in the development of a product and also the cost of any delay in the lead-time for its introduction to the company are both low. This is a state of happiness for all concerned and occurs very rarely. For this to be the case, a product area would need to be ring fenced with patent protection and any product guaranteed to be readily accepted by the market.

At the opposite extreme, the cost of lead-time delay is very high, the market may not be there when you arrive, and the product risk is high, it might not succeed in the market place. This situation is most common in rapidly changing high technology areas. It often arises when R&D, especially the strategic researchers, have come up with an innovative product or system which is outside the core competencies of the company. For instance, the chemical technology is not available within the company or simply the product is destined for a market where it does not operate. These are the occasions when a joint venture or licensing of the technology is considered (see Section C, 3.2 and 3.3).

A more common case is when the risk in developing the product is low but the cost of lead-time delay is high. Here, the market exists and the right product will sell, but the market share could drop or even disappear without a new product. In this case a crash programme of development is required. Typically this would happen when a company that has been operating successfully in a particular business area for a few years, has not appreciated that a fundamental shift is taking place, which will seriously harm its market position. A simple example in the Chemical Industry would be one where a manufacturer has been supplying a product as a solution, or dispersion, involving the use of an organic solvent. Products which are solvent free and easier to use in the workplace suddenly appear in the market. This is a consequence of not knowing what a customer really wants. A crash development programme is therefore required to avoid the disappearance of the company's business in this sector.

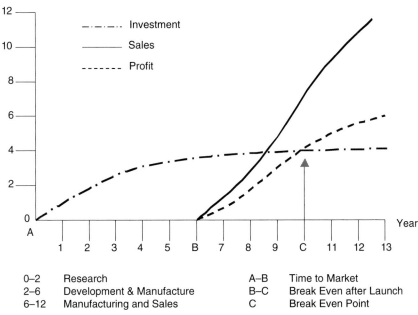

0–2	Research	A–B	Time to Market
2–6	Development & Manufacture	B–C	Break Even after Launch
6–12	Manufacturing and Sales	C	Break Even Point

Figure D14. The Return Map

If the cost of delay in lead-time is low it is usually because the product has a long lifetime once in the market; the customer once committed to the product will stay with it for a long time. If this is coupled with a high product risk, i.e. the product is expensive and needs a guaranteed customer, then a particular situation arises where it is best developed sequentially, as part of a series. An example outside the chemical field would be a range of aircraft such as the Boeing 747, 757, 767 series.

The crossover point in this diagram (Figure D15) is the area of medium cost of lead-time delay and medium product risk. In this instance it is usual to introduce the product in a modular fashion. Adding in improvements or sophistication as required by the customer.

2.3.4
R&D Management Issues

Speeding up the delivery of new products to the market involves intelligent risk and resource management.

The overall objectives of time compression management, within which an R&D Manager must work, are simply stated as "get products into the market more quickly". In order for this to be achievable the R&D Manager should have some overall targets set by senior management as part of the company strategy. At this level of management the targets do not need to be specific and can be as broad as:

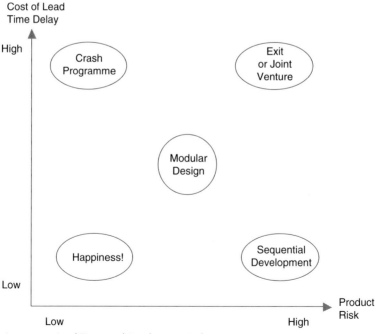

Figure D15.   Lead Times and Development Risks

- To reduce the time to commercialisation by 50%
- To have 25% of the sales turnover from new products by year XX

A Manager can then use these very broad targets in discussions with staff in their work groups. In this way their project and its objectives can be seen to fit in with the overall strategy of the company. The key element of the Manager's role is to ensure that the right environment exists for time compression in product development to be achieved.

It is generally accepted, that time compression management is best achieved through active collaboration between the members of cross functional teams comprising, as a minimum, R&D, manufacturing and marketing. The R&D Manager must play an active part in these teams, as they will not succeed if the contribution from their membership is half hearted. This is particularly true in the early stages, when the development plan for the new product is being set. At this time members of the team need to have a clear understanding of the objectives and commitment to the project, and of the contribution as individuals, they must make to achieve these objectives. In order to assist in successful time compression the R&D Manager should adhere to the following guidelines:

- Make sure that the targets are well defined, especially in terms of customer requirements. This has been covered in Section D, 1 and will not be discussed further at this point.
- Establish cross-functional teams at the start of the project. The team members must be people who are able to act on decisions made at project meetings, which should be action orientated and not talking shops.
- Ensure that time is the driving force not budgets. To achieve this make sure that non-value added activities are eliminated. Going faster on every topic will not be enough; there will be a need to eliminate the less fruitful activities. Whilst this is happening, quality or safety must not be compromised. The procedures inherent in the continuous improvement methodology of TQM should be employed wherever possible.
- Drive the process by setting an agreed date on which the product will be launched on the market. Since there is no guarantee of success in the research phase this may need to be modified in the light of progress. Certainly, once a product goes into development the launch date can be realistically set and adhered to by all parties. Have milestones that measure accountability. Only a major obstacle should be accepted as a reason for delay, whilst minor ones are there to be overcome. The use of tools, such as the Return Map, in deciding on greater expenditure to protect or reduce launch dates, should be encouraged.

2.3.5
The Interfaces between R&D, Manufacturing and Marketing

Most problems occur at interfaces.

Multi-functional project teams are used to overcome the problems that can occur at the interfaces between functions within the company. Even where such teams exist and are part of the culture, an R&D Manager should be aware that interfacial problems still exist, that they continue to operate outside the confines of the project team and can hinder its progress. Of particular interest to R&D Managers are the interfaces between R&D, manufacturing and marketing, particularly the latter. Having an understanding of the nature and origins of these problems will help Managers to ensure a smooth progression for the projects as they pass through the R&D phase.

One of the major reasons why interfacial problems arise is because the agendas and priorities for the three parties are different. The differences are illustrated in Table D1. The R&D Manager is always looking for opportunities to invent new products or processes. Both manufacturing and marketing recognise the importance of this activity to the future well-being of the company, but are working under today's pressures. In the case of manufacturing, it is under pressure to reduce costs and to deliver products into the warehouse that are required by marketing. Marketing is trying to meet the sales forecasts in the budgets and plans. These budgets will be mainly based on selling those products that are on the existing selling range. All three parties have a desire to improve the product range, the process, the output or supply and the service to customers, but only in R&D is the desire to change a positive driving force. R&D is a function where failure is part of everyday experience; it is the very nature of experimentation. For the other two functions, failure is at least very serious, if not disastrous, and is to be avoided at all costs. Innovative marketing is very desirable, but risks must be minimised, as failure must never be allowed to impinge on the customer-supplier relationship.

Successful R&D Managers are those who take account of these conflicting priorities, adjusting their approach to managers in the other functions as necessary. Coaching on these inter-functional differences should be given, preferably by the R&D Manager in person, to those members of a project team who are new to multi-functional team working.

2.3.6
The Overall Management of Time

It is over simplistic to think that by telling each Project team to improve their Time-to-Market that it will happen. In a multi-project environment each project team will be competing for resources and support to meet their personal goals. Conflicts between teams are bound to arise and it is imperative that management action be taken to avoid this happening.

An effective way to do this is to have a co-ordinating or steering group to oversee the portfolio of projects. This steering group will agree the overall priorities for the business, the goals of the project teams and arbitrate when conflicts arise during the lifetimes of the various projects. The management role of this group is to help, but not to interfere, with the day-to-day activities of the Project Managers, empowering them to deliver their projects as part of the strategy.

The size of this steering group is important, small but with the correct level of expertise and authority to gain the respect of the project teams. Ideally the group should have the following components:

Table D1   The behavioural differences between R&D, Manufacturing and Marketing

R&D	*Manufacturing*	*Marketing*
Devise new products	Keep costs down	Sell the existing range
Introduce a new product	Meet the schedule of production	Meet the sales forecasts
Improve the process for the product	Improve the output of the plant	Improve the supply to the customer
Change is desirable	Change is disruptive	Change must be planned
Failure is part of life	Failure is a disaster	Failure is serious

- Sponsor and Mentor. Usually and preferably a senior manager.
- Product (or Process) Champion. From the business, not a technical function. This may require more than one person if the projects cover more than one business.
- Technical Specialist. Broad experience of the area, possibly a technological gate-keeper, but not a narrow specialist.
- Project Managers. To attend only as required.

3
The Project Management Skills

Having dealt with the selection and evaluation of projects for R&D, the steps in the Innovation Chain and the importance of time compression management in the introduction of new products, the next topic to cover is the management of projects, with particular emphasis on the managerial skills that are required.

Of the many skills required in Project Management the following are the key ones:

- Team Selection and Management
- Project Definition
- Resource Allocation
- Methodology: Planning and Controlling

Another way of looking at this is to see it as a series of WHAT, WHEN, WHO and HOW questions to be answered:

- WHAT is to be delivered by the project?
- WHEN is it to be delivered?
- WHO is going to deliver?
- HOW is it going to be delivered?
- What CONSTRAINTS are there on the project?
- WHEN will they be placed?
- WHO will place the constraints?
- HOW will they be applied?
- What RESOURCES will be required?
- WHEN will they be required?
- WHO will supply the resources?
- HOW will they be supplied?
- WHAT will control the project?
- WHEN will control be applied?
- WHO will control the project?
- HOW will we know the project under CONTROL?

There are many commercial project management methodologies that cover these aspects of project management in different ways, and numerous texts on the topic and software, such as MS Project and its add-ons [D-13, D-14]. However, it is not the intention of this book to espouse the benefits of any one of these but rather to distil out the essential best practices from a range of such methodologies.

Research and Development in the Chemical and Pharmaceutical Industry, Third Edition. Peter Bamfield
Copyright © 2006 WILEY-VCH Verlag GmbH & Co. KGaA, Weinheim
ISBN: 3-527-31775-9

3.1
Project Manager and Team Selection

The nature of the project will define the managerial skill requirements of the leader.

After the evaluation phase, certain projects will have been chosen to go into the forward business plan and will need to be implemented. The first managerial task is to select the Project Manager or Leader and the members of the team.

Several points need to be considered and questions asked before the Project Manager is selected. The generic and first question that needs to be answered is "what type of project is it?" Does it, for instance, involve one or more of the following?

- The introduction of a new product or range of products
- The start of a new business venture
- The development and establishment of a new process
- The introduction of a new technology
- The design and commissioning of a new plant
- The carrying out of large scale customer or field trials

The nature of the project will define the functional or technical skills that will be required by the Project Manager. For instance, if the project involves the introduction of a new product, the person could be a chemist, biochemist or biologist from R&D or somebody from within the marketing function. For the design and commissioning of a new plant, an appropriate manager would be a chemical engineer from either R&D or manufacturing. The Project Manager for an extensive field or customer trial will be somebody with knowledge of the application technology, for example trials of a new agrochemical are very likely to be lead by a botanist or an ecologist.

Ideally the Project Manager would lead the project from its inception to completion. For projects under two years in duration this does not present a problem, but in the Chemical and Pharmaceutical Industry many projects take much longer than this to complete, especially new drugs and those involving new chemicals and chemistry. Typically these can take up to four years, and very much longer with a pharmaceutical product, from the start of research to release onto the market. Therefore we do not live in an ideal world and senior managers are faced with a practical reality; the time frame is often too long to keep one person involved as the Project Manager and likewise with the individual team members.

In these circumstances it will be necessary to change both Project Manager and probably team members during the course of the project. It is clearly better if this is done in a planned manner rather than having a change forced on the Manager by circumstances, such as an individual being promoted or moved to another function or business. One approach to this is to break the project down into management phases.

For example, let us consider a project that involves the introduction of a new speciality product. This is classically divided into the *invent-make-sell* phases. The process is much more complicated than this, and this has already been described in Section D, 2.2. Invent-make-sell is more accurately split into four phases, namely research-process develop-manufacture-launch. The initial project team should have representatives from the parties interested in all four phases, i.e. research, development, manufacturing and marketing, and these should remain represented throughout the lifetime of the project. A planned strategy for the change of Project Manager at some point during these phases can then be agreed. The use of two Project Managers is the best, and probably the only alternative to one Project Manager throughout the lifetime of the project. The use of three Project Managers is undesirable and the worst case of four should be avoided at all costs, since this is back to functional roles, making nonsense of the project team approach. As stated, the hand over point from one Project Manager to another should be agreed at the start of the project. A natural break can usually be identified based either on time or function and role. In this case it could be that the Project Manager for the research and the process development phase is from R&D whilst the manufacture and launch is managed by marketing personnel. The following is an interesting example of this two-leader/team approach [D-15].

> The Cibacron LS range of reactive dyes for cotton, sold by Ciba Speciality Chemicals, was developed in record time by a small team practising a new faster-to-market philosophy. In total it took two years to develop and introduce the new products, half the time it normally takes.
>
> Three people were involved in decision making during the project's first phase: one each from research, screening/application and marketing. After the research and selection phase, a second team took over the implementation and promotion: screening/application, marketing and promotion. Pre-testing was conducted with selected textile industry customers to fine-tune the product.

When involved in the choice of Project Manager for a multi-functional team, an R&D Manager needs to recognise that the driving force for completion of the project, as in the above example, is from manufacturing and marketing. These functions will be occupied with the detailed implementation of the outcome of R&D and are the best people to provide the management of the project beyond the research phase. This axis is illustrated in Figure D16.

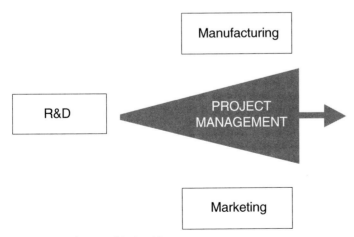

Figure D16. The Axis of Project Management

The Project Manager will need to have certain interpersonal and organisational skills to effectively run the team. In those companies, which are organised around project teams, the selection and training of such people will be of high priority. For those organisations where project teams are formed only as required, the R&D Manager will have to identify and foster such skills as part of the performance management process. Some of the key interpersonal skills required by the Project Manager are listed below.

- Planning
- Influencing
- Achieving
- Motivating
- Presenting
- Visionary
- Facilitating
- Delegating
- Negotiating
- Networking

It has been suggested that projects, in the category leading to a new business or new directions for the company, that are lead in the early stages by specific individuals who exhibit breakthrough creativity, as measured by the Myers-Briggs creativity index (see Section C, 1.5.1), show a much higher success rate [D-16]. However, this only applies if they also work within the framework of an excellent new product development system.

The individual team members will be selected for the technical and organisational contribution they can make to the project through its various stages. The nominations will come from the managers of the functions involved. The individuals should be good team workers; recognising the individual responsibilities within the team, contributing and behaving in a manner which supports the team and carrying out individual actions on behalf of the team. A list of the attributes demonstrated by a successful project team is as follows [D-17]:

- Commitment to the project
- Results-oriented attitude
- Creative thinking
- Willingness to change
- Concern for quality
- Ability to predict trends
- High involvement, interest and energy
- Capacity to resolve conflict quickly
- Good communications and feedback
- Mutual trust and confidence
- Interest in self-development
- Effective organisational interfacing

It is the responsibility of the R&D Manager to ensure that the people who are nominated to work on project teams fulfil these criteria and, if there is any doubt, deliver coaching or training on these aspects.

Assigning a person to a project team is in effect a secondment to that team. The individual's priorities and loyalties now lie with the project. For the period of the project, management of this resource is in the hands of the Project Manager. For individuals to feel happy in this situation they have to be reassured by the R&D Manager that the performance management and appraisal systems will take this twin management aspect and loyalty into account.

If the company is using a specific project management package, all members of the team must receive training in the principles and methodology of that particular package, before commencing any teamwork. This may seem to be obvious but it is easily forgotten, especially if for some reason a team member needs to be replaced at short notice.

The Project Manager will also have to demonstrate a facility with the hard skills required in project management; planning, estimating, budgeting, scheduling, tracking, controlling, measuring and re-engineering. In technical projects, it is advisable for the Project Manager to have a level of expertise that is recognised by the other members of the team.

3.2
Project Definition, the WHAT, WHEN and Constraints

Planning of a project should start only once it has been redefined in specific terms.

The Project Team Manager and members having been assigned to the project, their first task is to carry out a complete definition of the project. The team will know the overall goal of the project but its purpose and objectives will need to be clarified and redefined in more specific terms of deliverables, assumptions and constraints. Project definition enables all the stakeholders, both internal and external, to understand and agree upon the goals, responsibilities, assumptions and success criteria to be used. The planning of the project and the allocation of resources should only start once this has been done.

The redefinition of the project involves listing the deliverables of the project. These include the performance requirements of the product, process or service, the overall cost of the project and the time frame for its completion. Using these factors a statement can be written which defines the objective of the project. A typical example of a *project statement* would be:

> To develop and establish in two years time a new process for the manufacture of Product X, operating in a modified plant, to double the current production to 500 tonnes per year, the manufacturing cost to be 20% lower than at present, whilst the capital expenditure must not exceed 4 million euros.

This statement of the overall objective covers the WHAT, the WHEN and the CONSTRAINTS of the project.

WHAT
To develop and establish a new process for the manufacture of Product X to produce 500 tonnes per year at 20% less than the current cost.

WHEN
In two years time.
CONSTRAINT
The existing plant to be used and capital expenditure on modifications must not exceed 4 million euros.

Using this structured statement of its overall aims, the team is able to define the specific objectives and to refine and explore the other constraints of the project. A good way of looking at this is to consider what will have been delivered at the end of the project. At the same time any other objectives and constraints not covered in the general project statement, should be explored.

Applying this method to the example of the manufacture of Product X, the constraints are expanded to cover other aspects of HSE, personnel, time and money.

1. A new process for Product X
2. The ability to make 500 tonnes per year
3. The cost of manufacture 20% lower than previously
4. Use of clean technology with waste minimised
5. A safe process, with no impact on the work force
6. A process suitable for even larger scale manufacture
7. The potential for further cost reduction
8. The generation of intellectual property rights on the process
9. The work to be complete in under two years
10. The existing plant to be modified
11. The use of current plant items to be maximised
12. The capital expenditure on new items not to exceed 4 million euros
13. The disturbance to other manufactures to be minimised
14. The work to be completed with existing staff
15. The total revenue costs not to exceed 1,2 million euros

The Project Manager needs to check and agree the list of objectives with the sponsoring manager, and all other parties who are involved or have an interest in the project (the stakeholders).

Subsequently, the team will use the list of deliverables to determine the work to be done to achieve these goals. Several techniques can be used to carryout this process, the classical one is called the *work breakdown structure* (WBS), which was originally developed by the US defense establishment. "A work breakdown structure is a product-oriented family tree composed of hardware, software, services, data and facilities .... [it] displays and defines the product(s) to be developed and/or produced and relates the elements of work to be accomplished to each other and to the end product(s) [D-22]."

The work breakdown structure is therefore a method of describing the work to be done in the project in terms of deliverables and the tasks to be accomplished. It involves defining major deliverables and accomplishments during the project and listing a hierarchy of sub-deliverables and accomplishments. The deliverables and accomplishments are a sum of their sub components and the WBS is a sum of all the elements. A more detailed description of this can be found in any of the general books on Project Management [D-13].

Continuing with the Product X example, the headline activities in the WBS are:

1. Process development completed
    1.1 Laboratory process developed
        1.1.1 Alternatives routes evaluated
        1.1.2 Manufacturing suitability confirmed
        1.1.3 Costing done
        1.1.3 Product quality checked
    1.4 Hazard studies completed
        1.4.1 Chemical reaction hazards evaluated
        1.4.2 Plant operational hazards evaluated
    1.5 Patents filed
        1.5.1 Prior art checked
        1.5.2 Specification drafted
2. Plant modifications completed.
    2.1 Modifications designed
        2.1.1 Changes determined
        2.1.2 Alternative items identified
        2.1.3 Plant layout drawn
    2.2 Equipment obtained
        2.2.1 Suppliers evaluated
        2.2.2 Equipment ordered
    2.3 Plant items installed
        2.3.1 Pipe work installed
        2.3.2 Reactor installed
    2.4 Trials completed
        2.4.1 Water trials completed
        2.4.2 Instruments checked
3. Initial manufacture completed
    3.1 Process operated
        3.1.1 Plant Process written
        3.1.2 Computer controls modified
        3.1.3 Campaign carried out
        3.1.4 Cost and output rate estimated
    3.2 Planning complete
        3.2.1 Time slot identified
        3.2.2 Operatives trained
        3.2.3 Plant technical team chosen
    3.3 Materials delivered
        3.3.1 Materials ordered
        3.3.2 Quality checked
    3.4 Product cleared for sale
        3.4.1 Quality cleared
        3.4.2 Packed into sales containers

There are three main reasons for carrying out this part of the process:

1. To define the resources that will be required
2. To assign the responsibilities within the project team
3. To help in the planning stage

In drawing up a WBS care has to be taken not to make it just a long hierarchical list of things to do, including non-essential elements, so that the process becomes extended in time, which defeats the object.

3.3
Resource Allocation

The resources, which will be required in all projects, are human resources, support facilities, equipment and materials. Planning for these resources is the task of the Project Manager and the team.

- *Human Resource.* This is identified in terms of the technical contributions needed during the project's lifetime, the special skills or knowledge required from the people involved. How many people required, for what period of time and an estimate of the total cost needs to be made.
- *Support Facilities.* For example, a quantification of the cost of analytical support and other testing, such as HSE evaluation is made by the Project Manager. Any special work requiring additional development costs also needs to be identified.
- *Equipment.* The special equipment which may need to be hired and its cost listed. Consideration should be given to the need to use pilot plant facilities or large scale testing facilities.
- *Materials.* In addition to the bulk supplies required for the trial manufacture, the need for expensive or special quality chemicals for the laboratory considered and the likely cost.

This list or resource requirements, together with the activities generated during the analysis for the work breakdown structure, form the basis for the next phase of the process, namely the detailed planning of the project.

3.4
Planning, the WHO and the HOW

The detailed planning, by the Project Manager and the team, enables the tasks that have been identified, to be matched to the available resources, and organised within realistic time frames, in order to meet the agreed milestones or deadlines of the project.

The WHAT, WHEN and CONSTRAINTS have already been identified and the WHO and the HOW must now be defined.

Table D2   Prime Responsibilities in the Project Team

	Project Manager	*R&D Manager*	*Engineering Manager*	*Production Manager*	*Marketing Manager*
Process developed	Approve cost and quality	Develop Process	Provide Input	Provide Input	–
Plant modified	Approve capital spend	Provide Input	Modify plant	Provide Input	–
Initial manufacture completed	Monitor	Provide Input	Provide Input	Complete manufacture	Provide Input
Sales plan completed	Review	–	–	Provide Input	Develop plan

3.4.1
WHO

The human resourcing of the project defines WHO is actually going to carry out the WHAT of the project. The responsibilities for each aspect of the project are assigned to individual members of the team. This also allows any other responsibilities, which are outside their direct control, to be identified. On this basis the specific requirements of staff from each department or work group required for the programme to be a success will become clear. This definition and allocation of responsibilities can be illustrated in many ways. One of the simplest representations is a matrix, based on the activities identified in the work breakdown analysis and the functional responsibility within that activity (Table D2).

It is during this period of planning that the negotiating, persuasion and presentational skills of the Project Manager concerned are used to their fullest extent. This will be necessary in order to obtain the resources required, especially the human resource of the right quality. The R&D Manager, and all other affected managers, can help considerably in this process. They should be actively working with the Project Manager in the identification of the best available technical resource to meet the needs of the project.

It is at this point the revenue constraints placed on the project can be challenged, if they are thought to be limiting the eventual success of the project. Evidence will need to be generated to prove this point. The use of techniques like the Return Map can be used to justify any extra spending (Section D, 2.3.2).

3.4.2
HOW

This is a key step in the planning of the project. It is when plans are made, the order in which the activities are done and deliverables delivered, to fit in with the overall time scale of the project.

There are basic sets of tools, which are used to carry out this part of the process, all aimed at improving the overall management of the project. Most of these tools described below are available as individual items of software, or are part of the many software packages that have been developed for Project Management. If the company does not use one of these packages, a Manager should consult with the local technical software suppliers who should have a range of software available for purchase, and be able to provide the necessary advice on the scope and limitations of each package.

The four main methods that are used during the HOW phase in Project Management are:

1. Gantt or Bar Charts
2. Network Diagrams
3. Critical Paths
4. Program Evaluation and Review Technique

The first two, Gantt Charts and Network diagrams, are used during the detailed project planning phase whilst the second two are used to ask "what if" type questions about the derived project plans. An R&D Manager should aim to be familiar with these techniques. An understanding of these techniques is useful during the many debates that occur during the lifetime of projects. A brief description of each of these methods is given below.

3.4.2.1 Gantt Charts

A special type of bar chart used for providing a visual representation of the schedule of a project, bears the name of the inventor Henry L Gantt, hence *Gantt Charts*. In these charts the time frame of the project is represented on the horizontal axis in columns of days, months or years, whichever is appropriate, and the vertical axis has rows listing the tasks to be done during the project. Against each task a bar is drawn covering its duration.

	Operation	0–3	4–6	7–9	10–12	13–15	16–18	19–21	21–24
A	Process developed	▓	▓	▓					
B	Hazards Assessed				▓				
C	Plant designed				▓	▓			
D	Equipment ordered						▓		
E	Plant modified						▓	▓	
F	Chemicals delivered							▓	
G	Team trained							▓	
H	Trial completed								▓

Figure D17. Gantt Chart: Product X: Manufacture by the new Process in Modified Plant

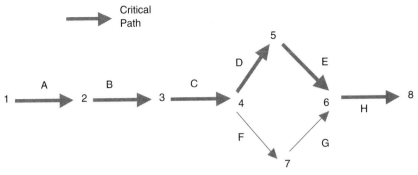

Figure D18. Network Diagram of Product X: Primary Tasks & Critical Path

A simplified Gantt Chart for the major tasks for the manufacture of Product X is shown in Figure D17.

Changing the time frame of any of the tasks on the Gantt Chart shows up areas where this will cause conflict and hence where priorities must be modified to fit in with the target completion date.

3.4.2.2 Network Diagrams

On the basis of events and activities and the order or sequence in which they will happen within the project the Project Manager can develop what is called a *network diagram*. The Project Manager also has to determine which activity takes precedence over another, i.e. which must happen before another can take place. These are, except in the simplest of cases, very difficult to construct by hand, unless the person is very experienced, and available software is of a great advantage.

Taking the list of tasks given in the Gantt Chart for Product X the order of precedence can be drawn up.

Task	Precedence
A. Process developed	None
B. Hazards evaluated	A
C. Plant modifications designed	A, B
D. Equipment delivered	C
E. Plant modified	C, D
F. Chemicals delivered.	A, B, C
G. Start up team trained	E
H. Trial Manufacture completed	E, F, and G

The network diagram derived from these activities is shown in Figure D18.

The complexity of the diagram rapidly increases when further tasks are added, particularly those that are carried out in parallel to the main activities. They can then be used for determining the critical path in a complex project.

3.4.2.3 **Critical Paths**

The *critical path* for a project is that which determines the shortest time in which a project can be completed. It is of greatest value where there are sophisticated network diagrams, involving multiple parallel activities.

The method involves the addition of the earliest start and latest finish time for each activity on the network diagram. Then using the list of precedents it is possible to calculate the shortest time in which the project can be completed.

> Adding up the times of each step in the Gantt Chart (Figure D17) of Project X gives a figure for the total time spent on these tasks of 30 months. However, because of overlapping activities the overall time is reduced to 24 months. Looking at the network diagrams in Figure D18 and the earliest start and latest finish dates in the Gantt Chart, it can be seen that F and G are off the critical path. Therefore the critical path follows A, B, C, D, E and H. It is these steps that must be looked at if the critical path and hence the time of the project is to be reduced.

When the times for each project task are very firm a special method, the Critical Path Method (CPM), can be used to ask more sophisticated questions about how to shorten the project time. The types of questions that can be asked are: "what are the benefits to the completion date by adding more manpower to task A or more capital into task B?"

3.4.2.4 **Program Evaluation and Review Technique (PERT)**

In the program evaluation and review technique (PERT), developed by the US Navy in the 1950s as part of the Polaris mobile submarine launched ballistic missile project, network charts depict task, duration, and dependency information. Each chart starts with an initiation node from which the first task, or tasks, originates. If multiple tasks begin at the same time, they are all started from the node or branch, or fork out from the starting point. Each task is represented by a line, which states its name or other identifier, its duration, the number of people assigned to it, and in some cases the initials of the personnel assigned. The other end of the task line is terminated by another node that identifies the start of another task, or the beginning of any slack time, that is, waiting time between tasks. Each task is connected to its successor tasks in this manner forming a network of nodes and connecting lines. The chart is complete when all final tasks come together at the completion node. A PERT chart may have multiple parallel or interconnecting networks of tasks. If the scheduled project has milestones, checkpoints, or review points (all of which are highly recommended in any project schedule), the PERT chart will note that all tasks up to that point terminate at the review node.

In order to construct a PERT chart the following need to be carried out.
1. Identify the specific activities and milestones
2. Determine the proper sequence of the activities
3. Construct a network diagram
4. Estimate the time for each activity
   - Optimistic time
   - Most likely time
   - Pessimistic

5. Determine the critical path

6. Update the PERT chart as the project progresses

The essential difference between PERT and CPM is that the former uses three types of time estimates as shown: the pessimistic, the realistic and the optimistic view.

The benefits of constructing a PERT chart are that it gives information on:
- Expected project completion time
- Probability of completion before a specified time
- The critical path activities that directly impact on the completion time
- The activities that have slack time and that can lend resources to the critical path
- Activity and start dates

It is beyond the scope of this book to deal with either the CPM of PERT methodologies in any detail and the reader should refer to specialist text- and websites on these topics.

3.5
The Formal Launch

A formal launch is like the opening night of a production.

When all the planning work has been done on the project it is time start work. It helps the Project team considerably if the project is formally launched. This can be done in a variety of ways but the essential element is that it be supported by the full weight of senior management.

This has parallels with the opening night of a new play in the theatre [D-17]. For example the "opening night" could take the form of a special meeting, involving the sponsor, held to announce the launch of a project. This can be followed by a memorandum, signed by the senior manager and the sponsor, to all departments and to people who are likely to be involved with or affected by the project, either as hard copy circular or an e-mail message or placed on an intranet bulletin board.

The importance of making staff aware that a project is underway, and that it is fully supported and being monitored by senior management should not be underestimated. The influence of a formal launch on the activities of both team members and other members of staff in the organisation is markedly positive.

3.6
Monitoring and Controlling the Project

The Project Manager has to have a good system for monitoring and controlling the project. Fortunately, most project management software packages contain such a system. However, slavishly following these systems, without a basic understanding of how they work, can lead to problems for the unwary. The basic tenets for these methods are:

- Having a clear knowledge of the criteria by which the performance of the project is going to be judged.
- Having a system for providing an early and accurate warning of any deviations from these performance criteria.

Control of the project is achieved by ensuring that all the tasks, or at least those which are on the critical path, are completed on time. For those elements of the project which have a relatively long time frame for completion, e.g. those which take months rather than weeks, the Project Manager may need to set up sub-projects in order to monitor the progress of these elements on a more even time basis.

When a problem is indicated by the control systems, Project Managers will need to bring all their skills into play in rapidly analysing the situation and making a decision on the corrective action to be taken. A good Project Manager will maintain an intimate contact with all aspects of the project and be able to anticipate problems and take avoiding action before they arise.

"What is an appropriate action in these problem cases?" The Project Manager can only formulate an answer to this question after discussions with the other members of the project team, and not in isolation. Any new action is likely to affect the plans for the tasks which come later in the project and team involvement helps to spot these and assess their impact. By this method full commitment to the agreed action will be gained from the team members.

Even in the best-planned projects, changes to the plans are often required due to influences beyond the control of the project team. If these non-controllable factors are fundamental, for example a policy change instigated by senior management, the project criteria will also need to be changed.

Major changes to the WHAT, at any time into the project, are disastrous. The target has been changed, in essence it is a new project and a new plan must be drawn up.

Changes to the WHEN or in the CONSTRAINTS are bad news to the project team, but can be resolved by modifying the existing plan, e.g. the work breakdown structure and resource planning need to be revised.

Serious problems can also arise from influences that happen without any prior warning.

> For example, during Project X there has been a fire in the chemical suppliers plant, making delivery of an essential material impossible at a late stage in the project.

These problems can be avoided by carrying out, during the planning phase, a potential problems analysis and possible emergency procedures. This involves the team analysing in detail each step in the project for problems that might happen to cause it to be delayed or disrupted, a risk analysis. Plans can then be made to overcome these problems or to have emergency procedures in place if they should happen; contingency planning.

> For instance, supply problems of raw materials in Project X was identified as a potentially serious problem. A second supplier of the essential material was identified and in addition an emergency stock of the crucial material was arranged.

Another key role for the Project Manager is monitoring the performance of the human resource responsible for carrying out the project tasks. The Project Manager's skills, which have been developed over the years through performance management, will need to be brought into play. The identification of the behavioural activities of individuals, which are detrimental to the success of the project, is an ongoing activity. The actions taken to influence an individual's behaviour must be carefully balanced against overall effectiveness in meeting the project goals and gaining the approval of the rest of the team. Heavy-handed treatment over a trivial example of inadequate performance can be counterproductive in achieving the goal of total team motivation.

Communications have an important part to play in the successful implementation of a project. External communications have already been discussed as part of the formal launch of the project. It is a good idea to have similar communications at important milestones during the project's lifetime. Internal to the project team it is necessary to have progress reports for team members and the human resource responding to them. To be effective the reports must be concise, accurate, and produced at regular intervals, with just sufficient detail for them to be speedily read. Progress reports, as well as notes on meetings, will also be required for inclusion in Project Folders as a part of the quality procedures within the company.

3.7
Completion of the Project

Project success needs to be acknowledged by senior management.

The completion of the project is followed in Project Management terminology by the *Closeout*. It represents the formal end of the project when the deliverables, the WHAT of the project have been delivered. Again, this should be widely communicated, and all those contributing to the success of the project formally thanked by the sponsoring senior manager. Formal reward is in order depending on the company's attitude to remuneration and performance related pay.

It is also the time to take stock formally of how well the project was implemented. The projects successes and problems are highlighted and critically reviewed in a positive way. The results of this review should be formally documented for future reference and made available for discussion within the company as part of the knowledge management system. This retrospective review of the project will make a serious contribution to the learning process of all Project Managers and team members, so that projects can be organised even more efficiently in the future.

References

[D-1] TROTT, P., *Innovation Management and New Product Development*, 2nd edn., Prentice Hall, Harlow, 2002

[D-2] KAPLAN, R. S., NORTON, D. P., *The Balanced Scorecard: Translating Strategy into Action*, Harvard Business School, Boston, 1996

[D-3] HAX, A. C., MAJILUF, N. S., *The Strategy Concept and Process: A Pragmatic Approach*, 2nd edn. Prentice Hall, New Jersey, 1996

[D-4] DAVIES, L., *Efficiency in Research, Development and Production – The Statistical Design and Analysis of Chemical Experiments*, The Royal Society of Chemistry, Cambridge, 1993

[D-5] *Environmental Performance 1994*, ICI Group Report. Millbank, London, 1994

[D-6] LANCASTER, M., *Green Chemistry*, The Royal Society of Chemistry, Cambridge, 2002

[D-7] *IBM Scientific Technical & Education Newsletter*, April 1995

[D-8] COOPER, R. G., *Winning at New Products: Accelerating the Process from Ideas to Launch*, 3rd edn., Perseus Books, Cambridge Massachusetts, 2001

[D-9] LEWIS, P. (ed.), *Health Protection from Chemicals in the Workplace*, Ellis Horwood, London, 1993

[D-10] SAXL, D., FAIRHEAD, J., Time to be the best, *Manufacturing Breakthrough*, March/April 1993, 21–24

[D-11] FLEMING P., Time for Change at 3M, *Research Technology International*, 1992, 17–19

[D-12] HOUSE, C. H., PRICE, R. L.,The Return Map, *Harvard Business Review*, January–February 1991, 92

[D-13] YOUNG, T., *The Handbook of Project Management*, Kogan Page, London, 1998

[D-14] BS6079, *Guide to Project Management*, BSI, UK, 1996

[D-15] Low salt Cibacron, *International Dyer*, August 1995, 10–11

[D-16] BURLEY, J., DIVINE, R., STEVENS, G., Creativity and Business Discipline = Higher Profits from New Product Development, *Journal of Product Innovation Management*, September, 1999

[D-17] BARKER, A., *The Alchemy of Innovation: Perspectives form the Leading Edge*, Spiro Press, London 2002

[D-18] BS 6079-1:2002 *Project Management. Guide to project management*

[D-19] CAVALLA, D., *Modern Strategy for Preclinical Pharmaceutical R&D*, Wiley, Chichester, 1997

[D-20] www.eu.int/comm/environment/chemicals/reach.htm

[D-21] European Chemicals Bureau, http://ecb.jrc.it/

[D-22] Military Standard (MIL-STD) 881B (25 Mar 93)

General References

Wherever possible in this book the references have been to works which are concerned with the activity of chemists and other professionals working in the chemical industry.

There are many excellent books on research management which are either non chemical or of a more general nature. The author has found several of these of value. A list of those consulted in recent years is included here, to assist in a wider understanding by the readership of this book.

[1] TWISS, B., *Managing Technological Innovation*. 4th Ed, Pitmans, London, 1992

[2] GILMAN, J. J., *Inventivity: The Art of Science and Research Management*, Von Nostrand Reinhold, New York, 1992

[3] GIBSON, J. E., *Managing Research and Developmen*, Wiley, New York, 1981

[4] TINGSTAD, J. E., *How to Manage the R&D Staff: A Looking Glass World*, AMACOM, 1991

[5] MILLER, D. B., *Managing Professionals in Research and Development*, Jossey-Bass Publishers, San Francisco, 1986

[6] JAIN, R. K., TRIANDIS, H. C., *Management of Research and Development Organizations: Managing the Unmanageable*, Wiley, New York, 1990

[7] KOCAOGLU, D. F., *Handbook of Engineering & Research & Development Management*, Wiley, New York, 1988

[8] BROWN, J. K., ELVERS L. M., *Research & Development; Key Issues for Management*, Conference BD, 1983

[9] DEAN, B. V., GOLDHAR, J. D., *Management of Research & Innovation*, Elsevier, Amsterdam, 1980

[10] BERGEN, S. A., *Project Management; An Introduction to Issues in Industrial Research & Development*, Blackwell, Oxford, 1986

[11] ROUSSEL ,P. A., SAAD, K. N., ERICKSON, T. J., *Third Generation R&D: Managing the Link to Corporate Strategy*, Harvard Business School Press, Boston, 1991

[12] TROTT, P., *Innovation Management and New Product Development*, 2nd edn., Prentice Hall, Harlow, 2002

[13] AINSWORTH, M., SMITH, N. I., *Managing for Innovation*, Management Books 2000 Ltd., Chalford, 1998

[14] GANGULY, A., *Business-driven Research & Development.*, Macmillan Business, London, 1999

[15] WHITE, R., JAMES, B., *The Outsourcing Manual*, Gower, Aldershot, 1996

[16] FUSFELD, H. I., *Industry's Future: Changing Patterns of Industrial Research*, American Chemical Society, Washington, 1994

[17] GELES, C., LINDECKER, G., MONTH, M., ROCHE, C., *Managing Science: Management for R&D Laboratories*, Wiley, New York, 2000

[18] MILLER, W, L., MORRIS, L., *Fourth Generation R&D: Managing Knowledge, Technology, and Innovation*, Wiley, New York, 1999

[19] LUDWIG, A., M., *The Price of Greatness: Resolving the Creativity and Madness Controversy*, Guilford, New York, 1995

[20] COHEN, C., M., COHEN, S., L., *Lab Dynamics: Management Skills for Scientists*, Cold Spring Harbor Laboratory Press, Cold Spring Harbor, 2005

Research and Development in the Chemical and Pharmaceutical Industry, Third Edition. Peter Bamfield
Copyright © 2006 WILEY-VCH Verlag GmbH & Co. KGaA, Weinheim
ISBN: 3-527-31775-9

Index

Research and Development in the Chemical and Pharmaceutical Industry, Third Edition. Peter Bamfield
Copyright © 2006 WILEY-VCH Verlag GmbH & Co. KGaA, Weinheim
ISBN: 3-527-31775-9